Laws of Nature

Interpretations

series editor: Roy Harris

This new series aims to explore the key concepts of intellectual enquiry in the Western world. Some of these concepts have a long and controversial history; others are of relatively recent origin. All are open to different interpretations by different thinkers.

Interpretations has two main aims:

to offer a survey of the main interpretations of an idea: those interpretations which have been influential in the history of Western thought

to offer experts an opportunity to present their own interpretation both of the idea and of its historical importance.

While giving the student an introduction to an important topic in the history of ideas, each volume in the series also represents an original contribution to that subject by a specialist in the field.

The language of the series is that of the general reader. All unfamiliar or technical terms are fully explained and critically evaluated.

Laws of Nature

Rom Harré

Duckworth

First published in 1993 by
Gerald Duckworth & Co. Ltd.
The Old Piano Factory
48 Hoxton Square, London N1 6PB
Tel: 071 729 5986
Fax: 071 729 0015

Distributed in USA by
Focus Information Group
PO Box 369
Newburyport, MA 01950

A catalogue record for this book is available
from the British Library

ISBN 0 7156 2464 4

Typeset by Ray Davies
Printed in Great Britain by
Redwood Press Limited, Melksham

Contents

Preface

In recent years the claims of science to offer an exclusive and dominant picture of the world have been called into question. On the one hand there has been scepticism about the value of the physical sciences as a model or pattern for investigations into central matters of concern in the way we live our lives. On the other, the scientific community has been revealed to be a society of all-too-human beings, intent as much on maximising their reputations as on pursuing the truth at all costs. The very idea that scientific truth is an attainable ideal has been questioned. And yet the achievements of the physical sciences are amongst the glories of human endeavour. These achievements are most compendiously represented in those principles we loosely call the 'laws of nature'. In this short book I have tried to bring out two main aspects of these laws through a discussion of a wide variety of philosophical reflections on them. There is the question of what the laws of nature are thought to be about, the subject matter and their content. And there is the question of how we come to assess their value as claims to knowledge. I have no final answer to offer to either of these questions. My aim has been to present some of the main ways that philosophers of science have thought about them.

Preface

I am grateful to the Delegates of the Oxford University Press for permission to use some short passages from my *Great Scientific Experiments*, and to the Editor of *Ratio* for permission to reprint some sections from my 'Dimensions of generality', vol. IV, pp. 25-33.

Linacre College, Oxford R.H.
& Georgetown University, Washington DC

1

A Sample of Laws of Nature

What is a law of nature?

When we think of a law of nature we think of a statement not just true usually or for the most part, but strictly and universally true. Nor do we take laws to be recordings of what happens perchance or for the most part, but specifications of what happens necessarily. To announce a law is to make a strong claim indeed. Are such implicit claims warranted? Do we really know the course of nature so well that we can claim to know her laws? The very phrase, 'a law of nature', is a metaphor, a survivor from a very different conception of the world from ours. For working scientists and most philosophers of science the phrase is almost completely secularised. Yet only about two hundred and fifty years ago Berkeley said: 'The set of rules or established methods, wherein the mind we depend on [that is the spiritual being who causes our ideas] excites in us the ideas of sense, are called the laws of nature, and these we learn by experience, which teaches us that such and such ideas are attended with such and such other ideas, in the ordinary course of things' (Berkeley, 1710: XXX).

Laws of nature were thought to be human versions of divine edicts, through which the divine governance of the world was accomplished. The metaphor of 'law' was close to a literal description. The great physicists of the sixteenth, seventeenth and eigtheenth centuries, such as Kepler, Galileo, Boyle, Newton and Descartes, were in different ways and in different degrees adherents to the doctrine of natural religion. Newton himself said, in his first letter to Bentley, 'I had an eye upon such principles as might work with considering men with a belief of a deity.' By studying nature, these men believed, we should be able to find out about God. The Creation should reveal much about its Creator. Two features of the laws of nature, as these were conceived by the natural religionists, have survived into the secularised concept. These are universality and necessity. But these features of what we now regard as the laws of nature will be in need of analysis and defence in their secularised versions. Can they be given a naturalistic interpretation? Before we can begin to answer this question, we must assemble a repertoire of examples of statements which are taken by everyone to be laws of nature, and see how they came to be discovered or conceived. We shall find that there seem to be various kinds of statements which are generally taken to be laws. It will be important to examine examples of each of these kinds of statements to try to discover what they are about. On this all-important matter philosophers have held very different views.

A catalogue of laws drawn from the writings of scientists, especially as these are summarised in textbooks, encompasses a great diversity of statements. I have chosen four main examples, each illustrating some facet of the generic notion of 'law'. Boyle's Law illustrates the use of an inductive procedure for the generalising of a mass of rather messy data culled from the manipulation of an apparatus. Snell's Law illustrates how experimentally derived laws have often turned out to be deducible from theory. Newton's First Law illustrates how a law came to be formulated through the analysis of concepts arrived at from thought experiments alone. It is a law which could never be instantiated, however subtle the apparatus we employ. Finally, Mendel's Laws will also illustrate a different process from that by which Boyle arrived at his law. Boyle abstracted his law from the data of experiment by an act of mind, so to say. Mendel abstracted his by a special experimental procedure.

The laws of nature, so called, are not the only statements of general import that we find in the discourse of the natural sciences. There are also those general statements with which a scientific community defines or prescribes what are to count as the types, kinds or species of beings that constitute its 'ontology', its world of interest. Classification systems, or taxonomies, are as important a part of the natural sciences as are the laws of nature. For instance, Mendeleyev's classification of the chemical elements, the Linnean system for classifying plants and animals, the classification of minerals used by geologists, the rapidly changing catalogues of subatomic particles play important parts in their respective sciences. Taxonomic systems are hierarchical: that is, individuals are collected under species, species under genera, genera under families, and so on. Each science has its own system of inter-related type-hierarchies. As we shall see, these play an extremely important part in the interpretation of the laws of nature. As well as analysing the structure and content of laws of nature I shall pay close attention to the kind of statement through which the principles of the taxonomies or ontologies of the various natural sciences are expressed.

Let us now turn to a detailed account of the origins and significance of the four laws I have chosen as examples upon which to base this book. The choice of a different quartet would have shaped the project in different ways.

Boyle's Law

Boyle's first set of experiments was designed to demonstrate the active power of the air. In his *New Experiments* (Boyle, 1660: 2) he says: 'Divers ways have been proposed to show both the Pressure of the Air, as the Atmosphere is a heavy Body, and that Air, especially when compressed by outward force, has a Spring that enables it to sustain or resist equal to that of as much of the atmosphere, as can come to bear against it, *and* also to show, that such Air as we live in, and is not condensed by any human or adventitious force, has not only a *resisting* Spring, but an active Spring (if I may so speak) in some measure, as when it distends a flaccid or breaks a full-blown bladder in our exhausted Receiver.'

But a more direct experiment was wanted. To demonstrate the active spring of the air as an observable phenomenon Boyle and Hooke set up an apparatus comprising a large tube from which the air could be extracted, with a smaller tube inside it. The smaller tube contained mercury which compressed some trapped air. While the outer tube was full of air the pressures balanced one another. But when the outer tube was evacuated the trapped air actively thrust out the mercury from the tube above it. I imagine that a quite spectacular fountain of mercury sprayed up out of the inner tube, as the air was suddenly extracted.

To complete the study of the air as a spring, Boyle proposed to make a 'measure of the Force of the Spring of the Air compressed and dilated': that is to measure accurately how the spring increased when the column of air was decreased by pressure, and how it decreased when the volume was increased by lowering the outside pressure. It was from this experimental project that his famous law emerged.

The experiment: measuring the spring of the air

The apparatus was relatively simple. Boyle and his assistant, Hooke, took a long glass tube 'crooked at the bottom' with 'the orifice of the shorter leg ... being hermetically sealed'. They carefully pasted strips of paper along each leg and marked them in inches. The tube was filled with mercury from the open longer end. Air was allowed to pass out from the closed end by 'frequently inclining the tube' so that 'the air in the enclosed tube should be of the same laxity as the rest of the air about it'. Then with the pressures equalised they began to pour mercury in the open end to increase the pressure on the enclosed air. They continued until the enclosed air was reduced to half its original volume.

By using what Boyle calls the Torricellian tube, a simple barometer, he and Hooke had measured the air pressure obtaining during the experiment. It was the equivalent of the weight of 29 inches of mercury. When the volume of enclosed air had been reduced to one half, the additional 'head' of mercury in the open end of the tube measured just 29 inches. In short 'this observation does both very well agree with and confirm our hypothesis... that the greater the weight is, that leans upon the air, the more forcible is its endeavour of dilation and consequently its power of resistance (as other springs are stronger when bent by greater weights).' At this point the tube broke. They tried again with a new tube of a 'pretty bigness'.

Using the newer, stronger tube they were able to make a series of observations examining the relation between the 'endeavour' of the air measured by the weight of mercury required to compress it, and the volume to which the original air had been reduced. The results are shown in Table 1.1.

A	*A*	*B*	*C*	*D*	*E*
48	12	00	Added to 22 1/8 makes	29 2/16	29 2/16
46	11 1/2	01 7/16		30 9/16	33 6/16
44	11	02 13/16		31 15/16	31 12/16
42	10 1/2	04 6/16		33 8/16	33 1/7
40	10	06 3/16		35 5/16	35 - -
38	9 1/2	07 14/16		37	36 15/19
36	9	10 2/16		39 5/16	38 7/8
34	8 1/2	12 8/16		41 10/16	41 2/17
32	8	15 1/16		44 3/16	43 11/16
30	7 1/2	17 15/16		47 1/16	46 3/5
28	7	21 3/16		50 5/16	50 - -
26	6 1/2	25 3/16		54 5/16	53 10/13
24	6	29 11/16		58 13/16	58 2/8
23	5 3/4	32 3/16		61 5/16	60 18/23
22	5 1/2	34 15/16		64 1/16	63 6/11
21	5 1/4	37 15/16		67 1/16	66 4/7
20	5	41 9/16		70 11/16	70 - -
19	4 3/4	45 - -		74 2/16	73 11/19
18	4 1/2	48 12/16		77 14/16	77 2/3
17	4 1/4	53 11/16		82 12/16	82 4/17
16	4	58 2/16		87 14/16	87 3/8
15	3 3/4	63 15/16		93 1/16	93 1/5
14	3 1/2	71 5/16		100 7/16	99 6/7
13	3 1/4	78 11/16		107 13/16	107 7/13
12	3	88 7/16		117 9/16	116 4/8

AA. The number of equal ſpaces in the ſhorter leg, that contained the ſame parcel of air diverſly extended.

B. The height of the mercurial cylinder in the longer leg, that compreſſed the air into thoſe dimenſions.

C. The height of the mercurial cylinder, that counterbalanced the preſſure of the atmoſphere.

D. The aggregate of the two laſt columns *B* and *C*, exhibiting the preſſure ſuſtained by the included air.

E. What that preſſure ſhould be according to the hypotheſis, that ſuppoſes the preſſures and expanſions to be in reciprocal proportion.

Table 1.1 Variation of volume and pressure

It is worth noticing that the experiment is not designed to discover what happens to air when it is subjected to a compressing force, but to find how the force exerted by the air is related to its state of compression. It is an attempt to measure the active power of air to resist force, its spring.

The experiment had been done at room temperature. What would be the effect of heating or cooling the trapped air? By putting a wet cloth around the tube they hoped to cool it, but 'it sometimes seemed a little to shrink, but not so manifestly that we dare build anything upon it'. However, when they cautiously heated the closed end with a candle flame, 'the heat had a more sensible operation'. The table recording these results includes figures that do not exactly conform to a law of strict proportionality. But, says Boyle, errors 'may probably enough be ascribed to such want of exactness as in such nice experiments is scarce avoidable'.

Boyle was very well aware of the problem of formulating universal hypotheses on the basis of a few experiments. 'But, for all that,' he says, 'I shall not venture to determine whether or no the intimated theory will hold universally and precisely.' 'No one perhaps yet knows how near to an infinite compression the air may be capable of, if the compressing force be competently increased.'

But the experiment as described tested only the effect of increasing the pressure on the air to greater than that produced by the weight of the atmosphere. There should be a corresponding reduction in pressure for air which has expanded beyond its normal volume. The apparatus had to be different, since they had no flexible tubes by which the surface of the mercury could be lowered. 'We provided,' says Boyle, 'a slender glass-pipe of about the bigness of a swan's quill.' They glued a paper strip with inches marked along it to the tube. The little tube was inserted into a wide tube, filled with mercury so that about one inch protruded above the surface. The pipe was sealed with wax to trap an inch of air within. 'After which the pipe was let alone for a while, that the air, dilated a little by the heat of the wax, might, upon refrigeration, be reduced to its wonted density.' By lifting up the slender pipe the air within was subjected to decreasing pressure, so that it was dilated to $1^1/_2$ inches, 2 inches and so on. They had already found that the barometric pressure was $29^3/_4$ inches that day, and to their satisfaction the difference between the levels of the tube when the air was dilated to double its original volume was only half the height of the barometer.

An error was found. When they replunged 'the pipe into the quicksilver' the air had slightly gained in volume at atmospheric pressure. Boyle supposed that this increase had come from 'little aerial bubbles in the quicksilver, contained in the pipe (so easy is it in a nice experiment to miss of exactness)'.

Let us refer again to Table 1.1 and compare the figure in the column marked E (which are the results calculated on the assumption that pressure varies inversely as volume) with the actual measurements recorded. We can see that the correspondence is not exact. However, it is close enough for us to suppose that the difference between a calculated result and the corresponding measured result is due to some inadequacy in the measuring process. Of course this assumes that the calculated result gives us the real state of the air.

Snell's Law

There was a delay of approximately two hundred years between the experiments performed by Boyle and Hooke and the theoretical demonstration of their law by Maxwell. By making certain assumptions about the nature of gases Maxwell was able to derive Boyle's Law and indeed all the major laws of the physics of gases. For my second example of a law of nature I have chosen Snell's Law of refraction. Only sixteen years separated his experimental discovery from its first theoretical derivation by Descartes. The law runs as follows: let *i* and *r* be the angles of incidence and of refraction of a ray of light passing through the boundary between two media (see Figure 1.1).

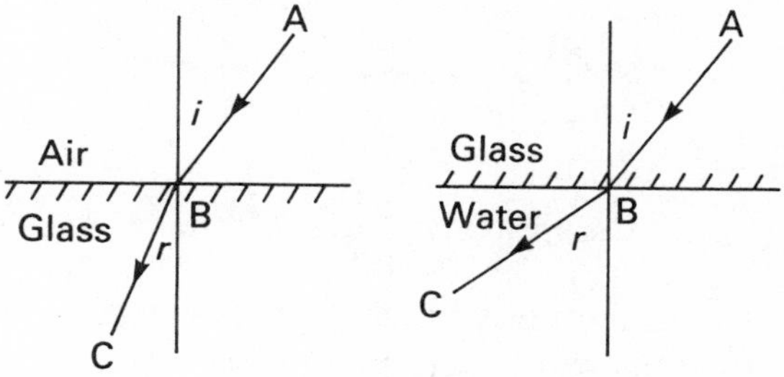

Figure 1.1 Refraction of a light ray at an interface between media of different densities. From a less dense to a more dense medium, the ray A, B, C is bent towards the normal. From a more dense to a less dense medium the ray is bent away from the normal.

Then, $\frac{\text{sine } i}{\text{sine } r}$ is a constant for each pair of media.

The fact of refraction is common knowledge. Most people know that the stone at the bottom of a clear pool looks nearer to the surface than it really is (see Figure 1.2).

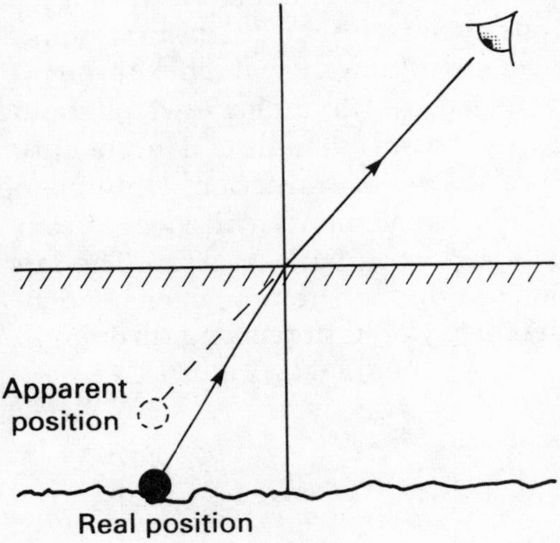

Figure 1.2 The apparent position of a stone in water.

Snell's achievement was to give a precise mathematical form to the law in such a way that it applied under all conditions and to all transparent media. According to Huyghens, Snell had arrived at this law in 1621 'after much effort and many experiments without, however, sufficiently understanding what he had discovered' (Sabra, 1981: 100). We do not know exactly how Snell made his discovery. His writings on refraction have been lost. All we have is his own expression of the law: '*radius versus ad apparantem in uno eademque medio diverso eandem habent inter se rationem. Ut secans complementi inclinationis in raro ad secanem [complementi] refracti in denso, ita radio apparens ad verum seu insidentiae radium.*' According to Sabra, this is equivalent to the formula cosec i : cosec $r = K$, where i and r are the angles of incidence and of refraction between the light ray and a line normal to the refracting surface.

In his *Dioptrique* Descartes (1637) provided an ingenious derivation of this law. First of all, he imagined what would happen in the case of a model, a mechanical analogue of the process of refraction. The model he chose was the glancing impact of a ball on a cloth. Here is how he describes the model: 'suppose that a ball driven from A to B [see Figure 1.3] encounters at the point B, not now the surface of the earth, but a cloth CBE, which is so weak and so thin that the ball can break it and pass entirely through, losing only part of its velocity ... . Only that part [of the motion] which would make the ball move from above downward can be changed in any way by encountering the cloth; the tendency which made it move toward the right should always remain the same as it had been, because the cloth is in no way opposed to it in that sense.' Suppose it loses half its speed in passing through the cloth. It will take twice as long to traverse BI as it took to traverse AB but its velocity along the cloth is unchanged. So, after the impact it will travel twice the distance AH in the horizontal: that is, as far as F in the time it takes to reach I. So, combining the two motions, horizontal and vertical, the ball will reach the point I where FE produced intersects the circle of radius BA. The ray is reflected when FE does not cut the circle AB. Descartes points out that, at a certain angle, a canon ball will bounce off a pond. People had been killed accidentally by virtue of that fact.

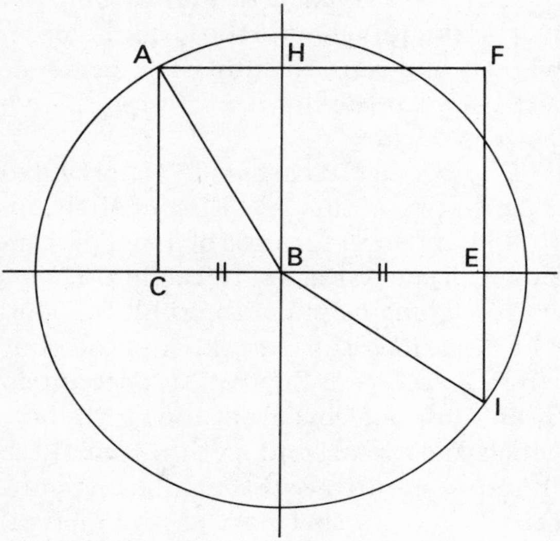

Figure 1.3 Descartes's mechanical model of refraction.
HF = 2AH; AB = BI.

Now let us imagine that the ball is struck by a racquet at point B. This will increase its speed to move more quickly on the far side of the cloth. If we work out the sums and draw the new diagram it turns out that the path of the ball will be deflected towards the normal. Now let us generalise this to light and the various pairs of transparent media in the interfaces between which refraction occurs. The ball represents the light corpuscle, the cloth represents the boundary between, say, air and water. The rays of light are more inclined in air than water to the surface where the refraction occurs. That is exactly the opposite of the case of the ball. Why? (See Figure 1.4.) Here is Des-

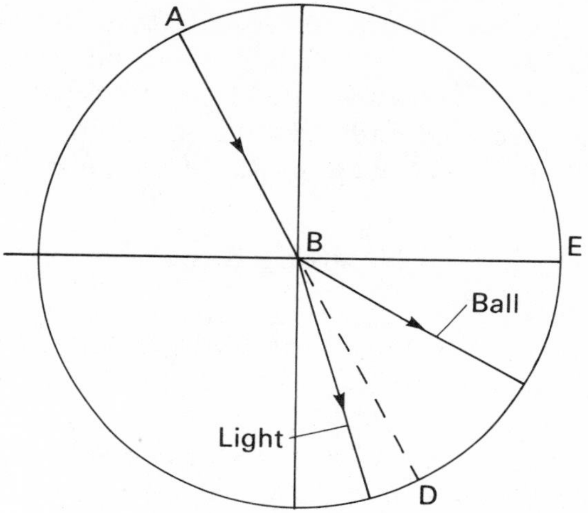

Figure 1.4 The path of light and the track of a ball.

cartes's explanation. 'Light is an action conceived in a very subtle matter. The action of this subtle matter may be very much more constrained by the parts of air which being as they are soft and loosely joined together do not offer much resistance to it. So much the small parts of the transparent body are harder and firmer, so much the more do they allow the light to pass more easily.' So, in the more dense medium, the light travels faster. It is as if it had been struck by a racquet at the surface. In the less dense medium, light travels more slowly. Descartes has given a theoretical derivation of the law. It is easy to see from the geometry of the diagram that the ratios between the sines of the angles of incidence and refraction are just as Snell said they would be. Subsequently, other theoretical derivations of the law have been proposed. Fermat based his derivation on his general principle that in any physical action the relevant particles will take the shortest path. Of course the proofs offered in the seventeenth and eighteenth centuries had to be reworked when it was realised that light should be taken as a wave motion for most of the purposes of geometrical optics.

Newton's Laws

For my third example, I take a law of the very greatest depth and generality: Newton's First Law of Motion. Roughly the law states that a body will continue in whatever state it currently is in, until it is acted on by a force. We must take it that all the material entities in the universe obey this law at all times and in all circumstances. However, we can never devise an experimental situation in which there are no external forces to act on a body. It follows that the law could never be demonstrated by means of experiment or observation. To appreciate the force of the law one must see it in the context of Galileo's analysis of the conditions of motion. Here is what Galileo (1638: 243) says: 'any velocity, once imparted to a moving body will be rigidly maintained as long as the external causes of acceleration or retardation are removed, a condition which is found only on horizontal planes; for, in the case of planes which slope downward, there is already present a cause of acceleration, while on planes sloping upward, there is retardation: from this it follows that motion along a horizontal plane is perpetual, for, if the velocity be uniform, it cannot be diminished or slackened, much less destroyed.'

In the *Principia*, Newton (1687) enunciates the law as follows: 'Every body continues in its state of rest, or of uniform motion in a right line, unless it is compelled to change that state by forces impressed thereon.' He adds some illustrations: 'Projectiles continue in their motions, so far as they are not retarded by the resistance of the air, or impelled downwards by the force of gravity. A top, its parts by their cohesion continually drawn aside from rectilinear motions, does not cease its rotation, otherwise than as it is retarded by the air. The greater bodies of the planets and comets, meeting with less resistance in freer spaces, preserve their motions both progressive and circular for a much longer time.' The law is clearly bound up with, and indeed depends on the notion of inertia.

The law itself as we have it in the *Principia* is closely related to two definitions that Newton offers of the concept of *inertia*. In the first, he says, '*vis insita* or innate force of matter is a power of resisting by which every body as much as in it lies perseveres in its present state whether it be at rest, or moving uniformly forward in a right line'. And then he adds: '*vis insita* may ... be called inertial force (*vis inertiae*) but a body exerts this force only when another force, impressed upon it, endeavours to change its condition.' Now clearly if we define material bodies as entities possessed of this *vis insita* Newton's First Law follows directly from these definitions. *Vis insita* is the power to persevere in a present state, whether at rest or in motion, and it is exerted only when another body tries to change its condition. So in the absence of that other body, that power simply drives the body forward or maintains it at rest.

Despite Galileo's powerful picture of sloping planes, it would be a mistake to think that Newton's First Law is drawn from experiments, or even such common phenomena as a puck sliding on the ice. To see this we have only to ask: 'Why does the puck keep going? Why does the space craft go on orbiting smoothly?' One might say: 'Well, all resistance has been removed, nothing is stopping it.' But does this mean that whatever was pushing it is now unobstructed? No, the point of the First Law is that in the case of inertial motion, nothing is pushing the mover. As Newton remarks, the inertial power of a material body only becomes active when an external force 'tries' to change the state of motion or rest. This extraordinary insight has another dramatic consequence. When a body is at rest, relative to some frame of reference, nothing is holding it in place either. From the point of view of the science of dynamics – that is the science of the causes of motion – rest and uniform motion in a right line are identical. So whether we say that something is moving or not will depend entirely on the reference frame to which we attend in determining the question. To an observer, the changes of the body α from point A to point B, could equally be due to the motion of the frame β or of the body, in the absence of the grander frame, to which we could refer (see Figure 1.5). Only the relative motion is real. But there is much more to be said about this deceptively simple proposition.

Newton's First Law occupies a commanding position among the laws of nature. In a certain sense, it serves to regulate the structure of the whole of physics. It is implicated in such fundamental questions as the status of space and time. Are space and time independent manifolds of places and moments at which material things exist, and events occur? Or are they just sets of relations between those things and events themselves? And yet the law seems so simple and even obvious. To open up the question of its true character, one can press the question touched on above. Is it necessary to act on a material body to keep it moving at a steady speed in a straight line? The obvious answer is 'Yes'. Without the horse, the cart would soon stop. Ship the oars and the boat will gradually cease to glide forward. But according to Newton's First Law the answer must be 'No'. In the light of the law, physicists generalised the concept of friction to the general idea of external resistances to motion. An effort is needed to keep a body moving only because its natural motion is impeded by some resistance. Without friction and other counteracting forces, a body would, it seems, move steadily in a straight line of its own accord or 'naturally'.

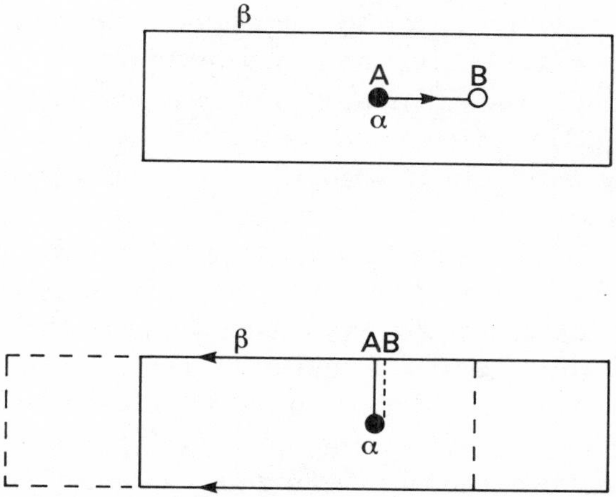

Figure 1.5 Two cases of relative motion.

The distinction between (caused) 'violent' and (uncaused) 'natural' motion can be found in the writings of the great Greek physicists. Aristotle, noting that bodies fell towards the centre of the earth of their own accord but needed to be pushed to keep going in a horizontal plane, identified the basic natural or, as we should now say, inertial motion with free fall. Motion in a straight line on the surface of the earth was, he thought, violent: that is, it was the result of impressed effort.

In a way Galileo and Newton and Gassendi simply reversed the attribution. Following Kepler, Newton assumed a law of universal gravitation, thus reclassifying free fall as a violent motion and, with his First Law, fixing horizontal motion in the absence of constraints as natural.

But how is Newton's First Law tied up with the problem of the status of space and time? Inertial motion in a straight line is natural motion, needing no impressed forces. Nor are forces needed to maintain a body in a state of rest on a horizontal plane.

So both inertial motion and rest are unforced or natural. How then could we tell whether a body was moving inertially or was at rest in empty space? We could look for some distant bodies, for instance the fixed stars, as a background against which to try to discern our motion. To make the question more interesting, let us suppose that there were no other bodies in the universe at all, and pose our question again. There are only two ways to tell whether a body is in motion. One is by seeing if it is being acted on by a force, the other is by referring its sequential positions to a frame of reference, a stable system of bodies, to which some convenient coordinate system has been attached. But in truly empty space there is no such system and by that hypothesis there are no forces acting on the body in question.

1. A Sample of Laws of Nature

The principle of Galilean relativity tells us that the laws of nature are the same for stationary and moving bodies, no matter what their uniform relative motions. So if we were confined to the closed cabin of a ship moving on still waters, we could not tell by experiment how fast we were moving, or indeed whether we were moving at all. We would need to open a porthole and look towards the shore to find a frame of reference. But in our imagined case there is no shore. Here is how Galileo introduces the principle of relativity in the *Dialogue concerning the two chief world systems* (Galileo, 1632 [1953: 186-7]):

> Shut yourself up with some friend in the main cabin below decks on some large ship, and have with you there some flies, butterflies, and other small flying animals. Have a large bowl of water with some fish in it; hang up a bottle that empties drop by drop into a wide vessel beneath it. With the ship standing still, observe carefully how the little animals fly with equal speeds to all sides of the cabin. The fish swim indifferently in all directions; the drops fall into the vessel underneath; and in throwing anything to your friend, you need throw it no more strongly in one direction than in another, the distances being equal; jumping up with your feet together, you pass equal spaces in every direction. When you have observed all these things carefully (though there is no doubt that when the ship is standing still everything must happen in this way), have the ship proceed with any speed you like, so long as the motion is uniform and not fluctuating this way and that. YOU WILL DISCOVER NOT THE LEAST CHANGE IN THE

EFFECTS NAMED, NOR COULD YOU TELL FROM ANY OF THEM WHETHER THE SHIP IS MOVING OR STANDING STILL.

Is this enough to show that in the physics of the Newtonian universe there is no place for the concept of absolute space and absolute time? Newton realised that these arguments did not apply to accelerated motions. They were never inertial. They always required forces and they always produced dynamical effects. This was the meaning of his Second Law ($F=ma$), that force is equal to mass-acceleration. Would it be possible to imagine experiments in which real accelerations could be detected in empty space by their effects rather than by reference to a frame of reference, such as the fixed stars, imagined absent in our thought experiment? Newton asks us to imagine two globes connected by a cord to which we can add a spring balance to check tension in the cord (see Figure 1.6). If the globes are rotating about their common centre of gravity, there should be a detectable tension in the cord varying with the rate of rotation. It should be zero when the globes are really at rest. These effects, Newton thought, would be observable to some disembodied observer, whether or not there were any other bodies in the universe.

Whatever we think would be the outcome of the imaginary experiment of the globes, the First Law stands untouched. It is an elegant expression of the principle of Galilean relativity. Straight line velocities can only be distinguished from each other and from a state of rest relative to some chosen frame of reference. The crucial point at the heart of Galileo's principle of relativity is that the laws of nature would always take the same form, whatever that relative velocity might be.

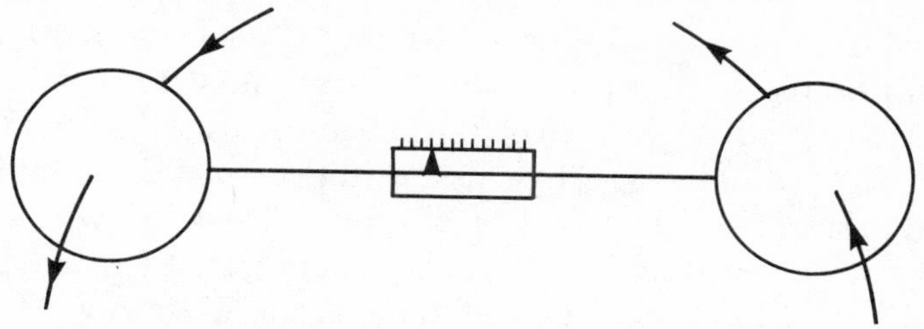

Figure 1.6 Rotating globes in an 'immense vacuum'.

Mendel's Laws

I suppose that among the best known laws of nature are Mendel's famous laws of inheritance. These laws depend on a distinction between two kinds of genetic factors, the dominant and the recessive. In any organism the inherited template, or genome, consists of pairs of genetic factors or genes. Each offspring receives one genetic factor or gene from each parent, reconstituting the full paired ensemble. The characteristics which that offspring will develop depend upon which of the inherited pair of genes is dominant and which recessive. To take a well-known example, which is not wholly accurate, from the genetics of human heredity: if the gene for brown eyes is dominant over the gene for blue eyes, then the offspring of one blue-eyed and one brown-eyed parent (with two dominant brown genes) will all be brown-eyed. However, each of these first-generation descendants will have a recessive blue gene. If any member of this group mates with the offspring of a similarly genetically endowed pair, the probability will be that the random shuffling of the genes will lead to a distribution in which one offspring in four will be likely to inherit the two blue or recessive genes, another one in every four will be likely to inherit two brown or dominant genes, and the remaining two will be likely to inherit one of each. These two will be brown-eyed since brown is dominant over blue. The statistical pattern of eye colour in the second descended generation then will be one blue: three brown (see Figure 1.7). With innumerable exceptions, variations and the like, these are the basic Mendelian laws.

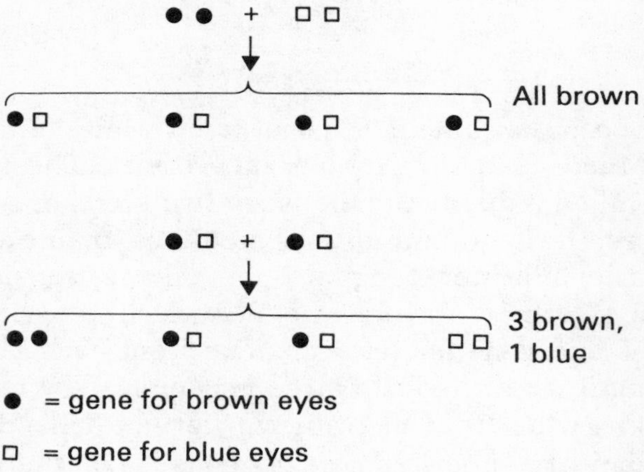

Figure 1.7 The genetic lottery in its simplest form.

How did Mendel arrive at these laws? Many attempts have been made to reproduce the famous experiments in which he studied the distributions, generation by generation, of four characteristics of peas: green and yellow colour and smooth and wrinkled skin. Fisher, the great statistician, thought that Mendel must have played fast and loose with the experimental results because the published figures describing them were 'far too good to be true'. We would expect any natural phenomenon to show a wide variation in the degree to which such strict arithmetical relationships would be found. We would have been very suspicious if Boyle and Hooke had reported experimental results that matched the results of calculation exactly. Fisher demonstrated that the chances of Mendel's getting results as good as those he reported from a patch of peas the size of a monastery garden would be about one in 30,000. Scholars have been puzzled as to what actually happened. Since Mendel's notebooks were destroyed by his successor, any reconstruction of his methods must be speculative.

A most interesting proposal has recently been made by Root Bernstein (1983). His analysis involves two very different levels of observation. Root Bernstein points out that in Mendel's time there were 'two fundamental ways of viewing nature'. The biological view saw individuals as varying continuously in all their common characteristics. Darwin shared this view and formulated his idea of the accumulation of small intergenerational differences as the explanation of the origin of species within that framework. The statistical view, on the other hand, was based on counting the frequency of occurrence of attributes in the population, but to be countable, attributes must be discrete. An individual must be either green or yellow if a head count of green peas and yellow peas is to be possible. To compare how many tall people there are with short, we must set a definite cut-off point which will allow us to treat the attributes tallness and shortness as distinct. Greenish-yellow things are an embarrassment if we are counting peas. In order to apply statistical methods to biology, it was necessary to solve, says Root Bernstein, the problem of assigning continuously variable characteristics to discrete categories. When we think about the basic ideas behind Mendel's theory of heredity, Root Bernstein seems to have described the situation very accurately. If there are genetic factors or genes, as we call them, and they segregate neatly according to the parental contributions, then surely they must be discrete, i.e. atomistic. On the other hand, if we line up a group of human beings, or examine a basket of peas, we find that they do not fall into two distinct groups, one containing nothing but tall people and the other nothing but short. There will be people having all sorts of intermediate heights between the tall and the short. And there will be many greenish-yellow and only slightly wrinkled peas.

There was then 'an irreconcilable tension' between the discrete categories of the theory of inheritance and the 'fuzzy biological' categories we need to describe plants and animals as they actually are. The 'reality of nature' confirms the 'ideology' of systems of classification, at least systems of classification that could yield materials capable of statistical treatment, that is, capable of being counted. Root Bernstein decided to reproduce Mendel's experimental programme not only by the planting and harvesting of peas but also in the way in which he imagined the results of the experiment might have been arrived at. Suppose that Mendel had instructed two of the novices to sort the peas. Root Bernstein tried just that. When we are faced with a bucket full of peas and two jars to throw them into, a jar for green peas and a jar for yellow peas, we find that about 7 per cent of the peas are not unambiguously classifiable as either green or yellow. We need a third jar to throw the borderline cases into. What do we do with the peas in the third jar at the end of the experiment? We have to *assign* them to the discrete categories embodied in the first two jars. According to Root Bernstein, 'what in fact Mendel published was not a "real" description of his peas, but his perception of how those peas could be categorised into "ideal" discrete groups'.

Root Bernstein presents very good evidence from his replication of the Mendelian experiments and from his study of some of the writings the abbot left behind, that Mendel was well aware of the need to make a decision about the unclassifiable peas, the outcome of which was not strictly determined by nature. In the end Fisher's criticism is beside the point: it would only be germane were there indeed such discrete categories in nature, but they are theoretical constructions by the help of which nature is segregated. As Root Bernstein says, 'Mendel's peas do not represent an objective truth' that Mendel could unambiguously interpret. '[They] represented a "set" of results, which required subjective analysis to fit into discrete categories. They were not data, i.e. given, they were *facta*, that is "made".' Root Bernstein's choice of the expression 'subjective' is misleading. Mendel's analysis depended on the use of *a priori* categories which emerged not from his experience of nature but from the exigencies of the method by which he had chosen to study the phenomena, that is, by counting the peas in such a way that the discrete categories did not absorb all the peas. He could have thrown away those which didn't fit. In Bernstein's replication the indeterminate cases are simply randomly assigned to the green and the yellow pots. The result is exactly Mendel's numbers, the numbers which Fisher thought were too good to be true.

The nature of the laws of nature

It should be obvious from these examples that the laws of nature are not simple descriptions of what can be seen to happen. They are descriptions of abstractions and idealisations from a somewhat messy reality. This observation gives rise to our first intuition. The question of what laws of nature are about may not be so easy to answer. Is the abstract and idealised law or laws about an abstract and idealised version of an actual process that the messy results record? Are Mendel's Laws about some kind of idealised pea plants growing in some fairy garden? On the other hand, is Boyle's Law about the real process which was actually going on when he and Hooke experimented, but which did not make itself manifest because of all sorts of interfering side-effects from other influences which were also active at the same time? There were fluctuations of temperature. And no doubt the apparatus shook with the passing of horses and carts in the High Street past Boyle's house, and so on. Now this query is as apposite for Boyle's Law as it is for Newton's First Law. What does Newton's Law describe? Certainly no motion that could happen in this universe, so full is it of impediments to motion.

Let us look first at the techniques by which laws as abstract and idealised propositions are extricated from the results of actual experiments.

All the examples I have discussed can be classified according to whether they involve an idealisation of a natural process or an abstraction from such a process. The distinction between idealisations and abstractions is somewhat artificial but it will prove helpful.

An idealisation preserves all the properties of its source but it possesses these properties in some ideal or perfect form. Boyle's Law applies to an ideal gas confined in an ideal apparatus, a gas which has pressure, volume and temperature just like an actual gas – only the pressure does not fluctuate. The inside of the tube is of uniform bore so the volume changes are regular, and so on. No such gas exists; nor does the apparatus with which we could study its properties. An abstraction, on the other hand, lacks certain features of its source. There are no greenish-yellow peas in the population of peas in Mendel's imaginary garden. His laws describe a process which is an abstraction from the kind of heredity process we can observe in complex organisms. There is no boundary between transparent media so perfect that the coefficient of refraction is exactly sine *i*/sine *r*. But we can imagine an idealised set-up in which the law does hold. Nowhere in the universe is there a space free of all impeding forces. Yet we can imagine a space, abstracted from all real spaces, in which motion is truly inertial. In our examples we have two cases of idealisation and two of abstraction.

We have also seen that there are two routes to the establishment of a law of nature: idealisation and abstraction from experiment, and theoretical derivation from some deep assumptions about the nature of the phenomena in question. It is easy enough if one has the requisite knowledge to follow a theoretical derivation of the law. It is not so easy to reproduce an experimental demonstration, as Boyle and Hooke found. It is not easy to bring people to believe in the results of an experimental programme unless they have an opportunity to reproduce the phenomena for themselves. A statement is hardly a bona fide law of nature unless the community can be shown how it is arrived at and how it can be applied in relevant situations. I shall draw on Gooding's (1989) analysis of how such public demonstrability is achieved.

In discussing Faraday's experiment demonstrating the principle of the electric motor Gooding addresses two main questions: How does someone like Faraday, working in the laboratory, produce a representation of the phenomenon in question in such a way that the role of the experimenter's guiding hand in producing the phenomenon is not revealed? That is, how does the experimenter make the phenomenon seem natural? The second question is related to the first: How does a physicist deal with novelty? That is, how is a new phenomenon made available to the scientific community who previously knew nothing of it? Gooding answers these questions according to two main principles, one sociological and one discursive. Knowledge representations of the right kind are produced by the interactions of various observers one with another in an interdependent group. But the process by which the group produces the relevant representation consists of a sequence of steps in which the phenomenon, originally a highly personal experience of its discoverer, is made publicly available to anyone with the cognitive resources to interpret that representation, and to manipulate an apparatus to produce it. An important part of this process occurs in the presentation of the results. If we consult Table 1.1 we can see that there are two columns of figures, one representing the measurements made with the help of the apparatus, and the other the calculations made with the help of the supposed law. In order to interpret the law, that the product of pressure and volume in a Boyle's Law apparatus is constant, we have to have the theoretical representation presented to us in such a way that we can see how to make an experimental representation of it. In the reconstructive process through which the method of production of the phenomenon is depersonalised and simplified, the complexity of the relation between an item of knowledge and its source is systematically concealed.

Not the least interesting part of Gooding's analysis is that he was able to show that there were three successive forms in which Faraday presented the method of producing continuous circular motion by electromagnetic induction. In Faraday's laboratory notebook 75 steps are recorded. In the first published paper presenting the discovery publicly there are only 45 steps. Anyone competent can follow these instructions and reproduce the phenomenon. Finally, there is the set of instructions Faraday published specifically to assist lay people to produce the effect. In that publication there are only 20 steps described. Similarly, when we learn Boyle's Law at school the cautionary tale that Boyle tells about the difficulties of collecting mercury from the floor of the laboratory after the tube has broken is omitted. In the schoolroom method the idiosyncrasies of Boyle's situation are left out. Why is there this difference in the presentation of a law? It is the role of the guiding hand of the actual experimenter which is written out of the successive accounts. The effect of this is to display the phenomenon as a natural occurrence existing independently of Faraday, Boyle, Snell, Mendel or anyone else. As we have seen dramatically in the case of Mendel's Laws, the technique by means of which the statistics for the laws were created involved eliminating the hidden hand of the assistants and Mendel's instructions to distribute the indeterminate peas at random. Root Bernstein had to imagine the concrete idiosyncratic reality of Mendel at work.

When Faraday made his original discovery the phenomenon of electrically produced mechanical motion was quite novel. How was it made available to the community as something they could see as natural? Gooding introduces a second main idea to analyse this process. He argues that in the reconstruction of the phenomenon various 'construals' are used. These may be ways of describing the effect, or they may be pictures or diagrams through which the phenomenon is presented iconically. The construals are not merely Faraday's or Boyle's or Snell's construals. As Gooding says, 'perception is shaped by the intention to communicate', and to communicate, in the first instance, with a certain, well-defined community, namely the other scientists working in allied fields. So an experimenter not only needs to work up the phenomenon for public presentation, but also for the more sophisticated and sceptical audience constituted by his or her colleagues. The reconstructed image in the Faraday case was 'motion in a perfect circle'; in the case of Mendel, the 'unbelievably precise' ratios of the distribution of inherited characteristics, and so on. Once the phenomenon has been made readily available to everyone interested both conceptually by construals and practically by reworking the experimental techniques, it can serve as the springboard for further research along the same lines. It is, to use Gooding's phrase, a 'heuristic'.

2

The Content of Some Laws of Nature

The problem of content

What, in general, are the laws of nature about? For instance, Newton's First Law: that every body continues in its state of rest or uniform motion in a straight line unless acted upon by an impressed force. What is this statement about? Is it about events? Is it about types or universals? Is it about dispositions? Or is it, in some peculiar way, about all three? Then, there is another question. Is it about what has actually happened or is it about what might, or could possibly happen? Is it about the actual events and states of the world or possible events and states? In the case of the Newtonian law just quoted, it seems quite obvious that that law must be about possible states of behaviour, of bodies, or kinds of bodies, since there is no such situation in which the law is actually exemplified. There are no bodies which are moving freely without the effect of some force or other upon their motion. Then there is a third question: rather more esoteric. Is a law of nature about everything in the universe or just about restricted groups of things? Many laws seem, at first sight, to be about everything. So when we enunciate Newton's First Law, we do not consider it as restrictive to any particular class of entities, and yet when we think of Mendel's laws it hardly occurs to us to ask whether they apply to planets, electrons, cities, oceanic waves and so on. So there are three questions about the content of a law of nature, i.e. what a law of nature might be about. Is it about events? or types? or dispositions? Is it about actual states of affairs, excluding the possible? Or is it about actual events and states? or possible events and states? or both? Finally, does a law of nature apply to everything?

Actualism: laws as summaries

We shall begin with a glance at the idea that the laws of nature are summary descriptions only of actual events. This is the Machean or positivist position. In a famous passage Mach (1894) offers the following account of the laws of nature. 'The communication of scientific knowledge always involves description, that is, a mimetic reproduction of facts in thought, the object of which is to replace and save the trouble of new experience. Again, to save the labour of instruction and acquisition, concise abridged description is sought. This is really all that natural laws are.' However it seems at first reading that many, or even most laws describe the behaviour and properties of substances and kinds of things that cannot easily be perceived. To conform to the Machean principles such laws must be reinterpreted. Mach gives an example (1894: 112-13): 'Let us picture to ourselves now two different electrical fluids, a positive and a negative fluid... in the fact that this conception reproduces lucidly and spontaneously all the data which arduous research only slowly and gradually discovered is contained its advantage and scientific value. With this too, its value is exhausted. We must not seek in nature for the two hypothetical fluids which we have added as simple mental adjuncts if we would not go astray.' So Mach offers us not only an account of the content of laws in terms of the patterns of actual events but also insists on a strongly reductive procedure to eliminate from our ontology all those hypothetical entities and substances with which we usually build theories to back up our descriptions of natural regularities. The electrical phenomena, in Mach's treatment, are such events as the bringing of a glass rod which has been rubbed with silk close to an electroscope. This action is followed by the leaves diverging. This is all, in Mach's view, that is involved in the law which we expressed in terms of positive and negative electricity. When we say that a positively charged rod was brought up to an electroscope and it attracted the electrons away from the leaves so that they became positively charged, and two positively charged material things repel one another, etc., this is only a picturesque way of describing the observed phenomena (see Figure 2.1). This is the classical, positivistic point of view.

It can come in two flavours. We usually believe that the world of science is the world we perceive: that is, the world of ordinary things, properties, their interactions and so on. And yet a theme in philosophy at least since the sixteenth century has been to cast doubt upon the easy assumption of the reality of that world. A more stringent form of positivism which is also to be found in the writings of Mach, involves a reductive treatment of material objects, the sorts of things that we usually believe exist independently of ourselves, to groups of sensations. Mach says (1894: 200), 'in men-

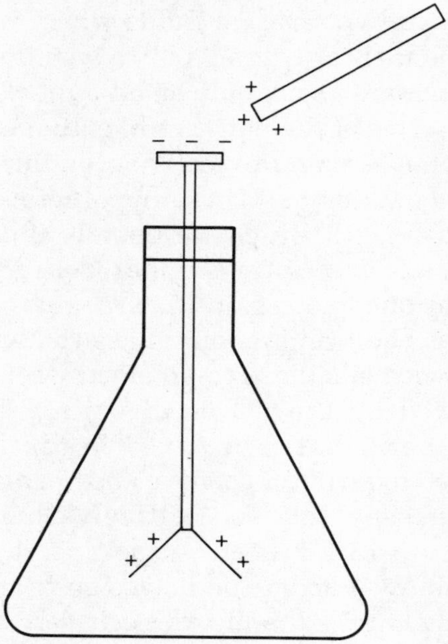

Figure 2.1 The behaviour of an electroscope.

tally separating a body and the changeable environment in which it moves, what we really do is to extricate a group of sensations on which our thoughts are fastened and which is of relatively greater stability than the others, from the stream of all our sensations.' Again (1894: 201), he says, 'it will be much better to say that bodies or things are compendious mental symbols for groups of sensations – symbols that do not exist outside the thought.' Now what becomes of the laws of nature? Well, at (1894: 205) we find this: 'The aim of research is the discovery of the equations which subsist between the elements of phenomena, and the elements of phenomena are sensations.' The laws of nature, then, are simply the compendious records of sensory regularities, correlations of elements. Any additional concepts which seem to refer to things in a world existing independently of ourselves are no more than symbols or devices for the compendious and convenient recording of general sensory patterns.

This view is very similar to that advocated by Hume. It has been very influential. In particular, it has the consequence that, in a general sense, what is recorded in the laws of nature is, from the point of view of logic, no more than a long-running accident. There can be no necessity in the pattern of our sensations since there are no real links between them. Furthermore, what is relevant according to the Machean view is only what is actual, what has happened and what will happen. What possible justification could we have on the basis of the Machean analysis of the laws of nature to extend their scope to what would or could have happened in the past and might or could happen in the future? Mach's position has sometimes been called 'actualism': the laws of nature cover only events which have actually occurred, or do actually or will actually occur. They do not comprehend possibilities. We are not justified in inferring a counterfactual statement from a law of nature. The law of descent seems to tell us not only how far bodies do actually fall when dropped, but also how far they would have fallen had they been dropped, and so on. The implausibility of this restriction will turn out to be of the greatest importance. One of the most potent reasons for thinking that the laws of nature are more than mere summaries of fact is that we generally assume, almost without reflecting on the matter, that if we believe a law of nature to be plausible then we can unproblematically infer all sorts of counterfactual statements from it. The law covers not only actual but possible cases.

Dispositionalism: laws as predictive schemata

If the laws of nature are propositions ascribing something to material things and substances – that is, if we are to justify some kind of anti-Machean position – what is it that they ascribe and to what do they ascribe it? For the most part, it seems the laws of nature are not about observable phenomena. Rather, they are used to ascribe dispositions, powers and tendencies to manifest certain appearances and observable or recordable states to material substances and experimental set-ups. Newton's First Law is used to ascribe a certain disposition to any material body whatever. Inertia is a disposition to resist changes to a body's state of motion or rest. Since inertia is a disposition we must say that bodies have inertia even when they are not resisting a change to their kinematic condition. The properties of material things with which the sciences deal are not the qualities we observe them to have, but the dispositions of those things to engender the states and qualities that we observe. When we speak of colour scientifically and ascribe a colour, say 'yellow', to a surface, we are ascribing to that surface a disposition to reflect light of a certain wavelength which, when the light interacts with a human sensory system, 'allows' us to see a yellow-hued expanse. We know that it has this power from the way it looks when illuminated, namely yellow. The word 'yellow' is of dual significance in physics. It has its everyday sense of the hue between orange and green, and it is also used to refer to that disposition of surfaces and transparent media to look yellow in the first sense. The surface or medium still has the same power or disposition even when it is in the dark. This is not an actualist interpretation of scientific laws. On this interpretation Snell's Law tells us how the passage of a ray from one medium to another *would go*. According to Snell the boundary between two transparent media of differing density has a certain disposition to deflect a ray of light in such a way that the angles of incidence and refraction will be related as they are in Snell's Law.

If we interpret laws of nature as ascriptions of dispositions to material things, any law so interpreted covers not only what actually happens but what would, could or might have happened in the past, as well as what could happen, or might happen, in the future. A disposition can be ascribed to a material thing, even when the relevant observable quality or state has never been actually instantiated. For example, the law of radioactive decay applies to those atoms which have decayed and to those which have not yet broken down, and even to those, given a sufficiently extended half-life, that will effectively never decay. One great advantage of treating laws of nature as the means for ascribing dispositional properties to material things and substances, including experimental set-ups, is that we escape the restrictions of actualism.

A further consequence of this interpretation is that it is no longer a mystery why a law of nature is ordinarily taken to imply a counterfactual statement such as the following: If a body continues to possess its powers and dispositions, then we can conclude that *were* this body, with these powers and dispositions, to be in such and such circumstances then the outcome would be the manifestation of the state, etc., specified in the disposition. In drawing 'counterfactual' conclusions we can be thought imaginatively to vary the conditions under which the substance, set-up etc. is manipulated or stimulated, while maintaining constant our conception of the nature of the being in question.

In recent years the concept of a disposition and how it can be used in interpreting the laws of nature have been elaborated in various ways (Cartwright, 1989). The most important development has been the introduction of the concept of an 'affordance' (Gibson, 1979). A knife can be ascribed all sorts of dispositions and powers, including its capacity or power to cut. This power depends, among other things, on the physical property of sharpness, the shape of the cross-section of the blade. 'Sharpness' could be manifested in all sorts of ways. Cutting is something people do with knives, a use to which a sharp knife is put. We shall say that a sharp knife affords cutting. What the knife affords is part of the meaning of a quite complex disposition.

If a knife is sharp, and if it is used in a certain way then it can cut whatever it is addressed to (within a certain range of materials). 'An affordance' is whatever a physical system can do in response to some human requirement. We shall speak elliptically of cutting as an affordance of sharp knives.

Instruments are like knives. They afford certain observable states and manipulative possibilities to an embodied human scientist, competent in their use. We could speak elliptically of a certain class of measurements as the affordances of an instrument, taking the requirements and skill of the scientific user of the instrument for granted. 'Transparent' instruments afford accurate representations of states of nature because we believe that, in the case of this class of instrument, determinate states of the physical world interact with the instrument to yield determinate and observable states of whatever physical system constitutes the instrument itself. We now know that there are some contexts in which the relevant instruments do not appear to be transparent. For example, the instruments used to detect or measure quantum phenomena are not transparent. A quantum mechanical 'state' of the world consists of an ensemble of wave states any one of which, with differing likelihoods, could be realised when the world and the instrument interact. The instrument affords a determinate state, but the state of the world from which that state of the world/instrument complex flowed was not determinate, at least as far as the resources of quantum mechanical description go.

Bohr's interpretation of this situation (Bohr, 1958) was to treat the laws of quantum mechanics not as descriptions of the world as it exists independent of the instruments used to study it, but as descriptions of the affordances of indissoluble instrument/world complexes. The subject matter or 'ontology' of fundamental physics shifts, in accordance with this interpretation, from the entities and processes of an independent world (which we could explore with 'transparent' instruments) to an ontology of instrument/world ensembles to which we can ascribe dispositions to afford this or that type of observation to a human scientist with a certain probability. In using instruments we do not manipulate the beings of a wholly independent world, but rather explore the affordances of a new kind of entity, the instrument/world complex, part of which we human beings have created. The Schrodinger *psi*-function does not describe an ensemble of wave-trains but a set of instrument/world affordances.

Laws as statements about universals

There is a third way in which the laws of nature could be interpreted. It has been less popular in recent years than the extentionalist and dispositionalist interpretations. It has been advocated by Dretsky (1977) and Tooley (1987). It has taken on a new lease of life with the type hierarchy theory of laws of nature proposed by Aronson (1992) and Way (1991). The idea is a very simple one: laws of nature relate properties to properties. How the properties referred to by a law of nature are related is the key question. One way is as subtypes in a type hierarchy. To say that (lawfully) all acids have free hydrogen ions is to say that type 'acid' is a subtype of the supertype 'substance with free hydrogen ions'. Another way would be to interpret the law statement as declaring that the two related properties, say 'ravenhood' and 'blackness' are both manifestations of the same underlying avian nature, that is, whatever it is to be a raven. Laws of nature are not compendious summaries of facts, as Mach would have it. Rather, they express the relations we believe to obtain between 'universals'. Laws of nature tells us a huge variety of different kinds of things, but the general structure of the laws of nature is always a relation of kind to kind. According to this view the content of any law of nature is best understood as simply a snippet or cutout of a type hierarchy. It turns out that this interpretation can naturally and fruitfully be combined with the more traditional dispositionalist view.

For example, Newton's First Law relates as subtype to supertype beings of the material entity type to beings characterised by belonging to the type which affords inertial motion. Snell's Law relates certain types of interfaces between media to an abstract characterisation of a certain type of geometrical conformation of rays of light. And so on. The types in question are characterised by their affordances, since first and foremost they are types for the human manipulator or explorer in his or her role as scientist. The types relevant for science are not characterised by their appearances but by the dispositions they possess, either independently or in indissoluble interaction with apparatus, to appear in certain ways to human observers.

Poincaré and the redescription theory

There is quite another way of understanding or interpreting laws of nature which we owe to Henri Poincaré (1958). The basic idea in his analysis of science is that we should make a distinction between 'brute facts' and 'scientific facts'. A scientific fact is simply a brute fact redescribed in a scientific language. The result of the redescription is another statement, with the same content though it will be less specific. I look out from my window and see a sunset. As described in ordinary language it has the character of a brute fact. If an astronomer redescribes the sunset in terms of the heliocentric theory of the solar system he or she must draw on another language, the language of astronomy. So redescribed, it would refer to the occlusion of the sun by the rim of the turning earth. Or it might be redescribed in yet another language, the language of geocentric astronomy. The same brute fact would now be expressed in the statement 'the moving sun has dropped below the rim of the fixed earth'. In the laboratory, when faced with an electrical circuit, we might ask: Is a current passing? What might be meant by that query, according to Poincaré, is something like this: Will the ammeter change its state in such and such a way? But it might also mean: Will the voltmeter, appropriately connected, change its state in a certain way? Both of these questions are redescriptions in the language of physics of a query about whatever brute fact it is that is true of the electrical circuit. Now what is the point of the law which links current and potential difference '$V=IR$'? If we are looking at two redescriptions of the very same fact, the only possible role for this law can be as a rule of synonymy for a pair of terms of the scientific language. That is, instead of thinking that α effects are accompanied by β effects, we should more properly understand the law as saying that the expression 'V' is synonymous with the expression 'IR', that is, where the one appears we could write the other, preserving the brute fact for which they are alternative descriptions. Each scientific language will have its own rules of synonymy.

The relation between rival or successive theories proposed for the same range of phenomena should be treated in the same way. The difference between the Copernican and the Ptolemaic astronomy is not, according to Poincaré, truly a difference between two different visions of the universe. Rather it is the sort of difference that we might find between French and

German. It is a difference between languages. So if we translate the word 'sunset' into French, we get '*coucher de soleil*'. But if we were to translate it into German, we would use the word '*Sonnenuntergang*' to express the same brute fact. Since these are both descriptions of the same brute fact, there is a translation rule which links the French word with the German. Within the languages there are of course synonymy rules which link one German word with another, one French word with another, one English word with another and so on. These correspond to the laws of nature as expressed in the languages of science.

What are we to make of the possibility that further investigations of some field of scientific interest may turn up counterinstances to our original hypothetical law? How could that be if the laws of nature are the rules of synonymy for some scientific language? Poincaré answers that falsification can always be countered by the elevation of the rule of synonymy to the status of a definition. This he calls creating a principle. Suppose 'A' and 'B' are scientific terms related by synonymy and we find that the phenomena, let us say, of current and voltage are not always uniformly related, as the law $V=IR$ would have it. Falsification is avoided by introducing a fictitious entity, 'resistance', which is related by synonymy to some other expression, say 'temperature', in such a way that the original law is preserved. We now have another terminological relationship which is open to further revision, by a repetition of the defensive procedure. This way of interpreting the laws of nature, particularly the laws of physics, has come to be called 'conventionalism'.

In summary, Poincaré's ideas amount to this: the work of scientists is strongly directed to creating conceptual structures out of concepts which are internally related to one another. These internal relations can be at the level of adjustable rules of synonymy or they can be elevated into rigid principles. Laws of nature are necessary truths, that is rules of synonymy. But they are neither arbitrary nor *a priori*. We do not invent whatever rules of synonymy we like. Nor, of course, do we approach nature with them ready-made. We try to construct a language within which the brute facts of experience are expressed as comprehensively and as elegantly as possible. The job of science is the forging of a language precisely suited to that purpose.

Which of these ways of interpreting the laws of nature is right? We shall see that all four have their place in a comprehensive account of the variety of statements that have some claim to be counted as laws of nature. Each interpretation highlights a certain aspect of the jobs for which the laws of nature serve as conceptual tools.

3

Generality

The scope of laws of nature

Laws of nature, it would seem, should be general in their application. But nearly every law of nature is hedged around with restrictions of one sort or another. In particular there are usually, often explicit, boundaries set to the range of cases a law is expected to apply to. To explore these matters I shall begin with the most obvious ways in which the generality of laws is restricted. Let us represent some sample law of nature by '*L*'. For laws of least restricted scope, such as Newton's First Law, we have something like the following:

> For all substances; and for all experimental conditions; and for all space-time locations *L* holds good.

I shall call each qualification a 'dimension' of generality. There are three implicit dimensions of generality for most laws. Unless qualified, they are assumed to have 'substance-generality', 'experimental-generality' and 'spatio-temporal-generality'.

In the natural sciences very few law-statements are not subject to restrictions of scope. In practice laws are qualified in each dimension of generality. So we have something like the following:

> For all substances of a certain kind; and for all experimental conditions within a certain range; and for all spatio-temporal locations within a certain region *L* holds.

I shall call such qualifications 'substance-restrictions', 'experimental-restrictions', and 'spatio-temporal-restrictions' respectively. In general, such restrictions are forced upon us by the discovery of counter-instances to our tentatively proposed law. We shall see that the effect of the discovery of a 'counter-instance' to a law on our understanding of that law is different in each dimension of generality. We shall also see that there is a pattern to the transition from 'optimistic' and unrestricted law-statements of the first type to restricted law-statements of the second type above, as discoveries of the limits to the application of a tentative law are incorporated in it. This pattern is distorted by the formulation of the logic of the transition in Aristotelian form, which would look something like this:

> From the truth of 'Some S is not P' (exceptions to the law 'All S are P' discovered in an experimental programme) we infer the falsity of 'All S are P'.

The use of Aristotelian logic to express the pattern of reasoning that follows the discovery of a limit to the application of a law suggests the following picture of scientific advance, a picture proposed by K.R. Popper (1959). We start with Boyle's original law in unrestricted form.

> All gases obey *PV=k* under all conditions (hypothesis).
>
> Some gases do not obey *PV=k* under certain conditions (experimentally discovered counter-instance).

We are forced to conclude that our original hypothesis, that all gases obey *PV=k* under all conditions, is false. False hypotheses should be rejected, so we are back at the beginning again. All we can do if confined to the moves of Aristotelian logic is make another conjecture. Try:

> 'All gases obey *P(V–b)=m* under all conditions (new hypothesis).
>
> Some gases do not obey *P(V–b)=m* under certain conditions (experimentally discovered counter-instance).

We are now forced to conclude that our new hypothesis is also false. All that we now do is to proceed to try another conjecture.

As an outline sketch of scientific method this picture has a certain degree of plausibility. But it fails to give any rationale for the 'Try'-moves. In real science each new hypothesis in a research programme is related to its predecessor in such a way as to incorporate the evidence that led us to reject that predecessor. These moves are not to be dismissed as intuitive, a-logical or irrational.

The relations which, in fact, obtain between rejected hypotheses and those proposed to replace them are, on the surface, very complicated. By attending to the differences between the three dimensions of generality we shall find, in each dimension, a systematic pattern of reasoning, a 'logic' if you will, in which the way new hypotheses are arrived at can be rationalised. However in each dimension of generality this 'logic' is different. To find these systematic patterns of reasoning we need to notice that both before and after restrictions have been placed on the scope of a law in each dimension of generality there is still a potential infinity of cases within the restriction. For instance, if we restrict the application of a law to the surface of the earth there are still infinitely many places on that surface at which we presume the law will still hold. Similarly restrictions on substance-generality or experimental-generality still leave infinitely many specimens of the permitted substances, and infinitely many experimental conditions empirically unexplored, at least in principle. All this, of course, requires assumptions about the mathematical properties of the spatial and temporal manifolds and about the continuity of variation in physical properties like pressure and temperature.

The management of evidence that contradicts a supposed law

Taxonomic and non-taxonomic generalisations

We must also notice that the effect of a counter-instance is determined by which dimension of generality it restricts. The difference in these effects can be brought out if we adopt another way of classifying laws of nature:

3. Generality

(a) Taxonomic	(b) Non-Taxonomic
(i) Potentially analytic	(i) Disguised identities
(ii) Non-potentially analytic	(ii) Non-identities

For this classification we need the distinction between the intension and the extension of a class-term, like 'sheep'. The intension of the term (or sometimes we say of the class) is the set of properties which a being must have to count as a member of the class, to fall under the term. To be a sheep a being must be an animal, a quadruped, an ungulate, woolly, etc. The extension of the class-term (or the class) is the set of actual beings properly so called: that is, all the sheep. It is easy to see that the more properties there are in the intension, the more demanding it is as a criterion for class membership, so the fewer beings there will be in the extension of that class. Intension varies inversely as extension.

I shall say a generalisation is 'taxonomic' when the order of the members of the extensions of the general terms in the generalisation is not relevant to the import of the generalisation, and non-taxonomic when the order is relevant. It does not matter in which order we examine the sheep, but it does matter in which range of temperature we study the behaviour of gases.

Taxonomic generalisations are characteristically found in systems of classification. For example the Linnaean classification of organisms depends upon the acceptance of statements delineating the characteristics proper to species, genera, etc. The classification of stars depends upon the acceptance of statements delineating the characteristics of different kinds of stars: binary, quasars, red giants, white dwarfs and so on. The classification of the chemical elements depends upon the acceptance of statements delineating the characteristic reactions which each element exhibits. A necessary requirement for the use of such general statements as the bases of classificatory schemes is that the members of the extension of their terms can be taken in any order. If we depend upon the acceptance of general statements like 'All mammals suckle their young', 'All mammals are warm-blooded' and so on to demarcate a family of vertebrates, the properties or characteristic behaviour referred to by these general statements must be attributable to particular mammals irrespective of the order in which these instances are examined. Only thus would we have a classificatory system: that is, a scheme which can be used to place individuals, wherever and whenever found, into their appropriate subclasses.

Within the class of taxonomic generalisations I suggest we distinguish those which are potentially analytic from those which are not. The potentially analytic are those generalisations which are, have, or might be employed in a formal definition of a biological species or chemical element. For instance, the generalisation 'All mammals suckle their young' is potentially analytic, for we could use it as a definition of the class 'mammal', in which case 'suckle their young' becomes part of the meaning of the word 'mammal'. It may also be generally true of mammals that their habitats range from 1000m below sea-level (whales) to 5000m above (ibex). But it would be odd indeed if a warm-blooded, young-suckling vertebrate were refused the classification *mammal* because it was found on the upper slopes of Everest. Of course 'potential analyticity' is relative to the classification scheme which is actually in use, for we could choose to have a quite different classification of organisms in which families like 'bammals' figure, defined by the maintenance of a constant temperature irrespective of their reproductive biology. 'Suckle their young' would be irrelevant to their classification, just as colour of pelt is usually irrelevant in our present classificatory system, though not in that used by Australian Aboriginals. The distinction between general statements which are 'potentially analytic' and those which are not is not an absolute logical categorisation which could be determined by analysis and inspection of isolated general statements. Nevertheless it is readily recognised for any given classification system.

Not only can we talk of taxonomic generalisations, but we can also use the notion of general terms having taxonomic generality. I shall say that a general term has taxonomic generality when it is employed in such a way that no order of the members of its extension is presupposed. The quantifiers of ordinary language – namely 'all', 'each', 'every', 'any' and so on – are ambiguous, since they could be used in either taxonomic or non-taxonomic generalisations. But the quantifier 'any' is unambiguously taxonomic, since it is used explicitly to rule out any presupposition of order in the members of the range of individuals quantified. Any material body will continue in its state of rest or uniform motion unless it is acted upon by an impressed force.

A non-taxonomic generalisation is one in which order in the extension of the general terms is presupposed. Such generalisations are almost always arrived at through the examination of an ordered set of cases. Non-taxonomic generalisations are characteristically found in those branches of science where functional relationships, for instance mutual variation, is the characteristic form of a law of nature. Wherever the members of the extension of terms are numerical we have a built-in ordering principle, since the members of the extension of the independent variable can always be ordered numerically. Boyle's Law is just such a generalisation. Once they are ordered, the extensions of the dependent variable or variables are ordered by the functional relation that holds with the independent variables. Indeed so close is the logical liaison between functional laws and ordered sets that we can, in mathematics, use the notion of ordered sets of pairs, triads, etc. to define the concept of 'function', just as an ordered set of pairs, triads etc.

The subclassification of non-taxonomic generalisations into disguised identities arises in the following way: there are many cases where we express identity by a formula which looks just like a genuine functional relationship. For instance in physics we are presented with an apparent functional relationship between the temperature of a sample of gas and the average kinetic energy of the molecules in the sample. It looks at first sight as if we are learning of a relationship that holds between two independent physical characteristics of a gas, the temperature of a sample and the average kinetic energy of its molecules. But reflection shows that this is not the case, since 'temperature' *is* 'mean kinetic energy of the molecules'. What is presented as a functional relationship is, when interpreted within the conceptual system expressed in the molecular theory of gases, an identity. Identities like this are mainly used to link theoretical models or mechanisms to observations. In contrast a functional relationship between the pressure, volume and absolute temperature of a sample of a gas is not an identity since we are relating by that function physical characteristics of gases that are independent in a certain sense. Their independence is shown by the fact that what is expressed by the tri-variation $PV \alpha T$ can be expressed by three bivariations between P and V, V and T, and P and T.

The Square of Opposition

In classical logic a proposition can be either true or false. If it is true it is not false, and if false it is not true. Is the elementary logic of truth and falsity the best way of expressing our assessments of the worth of scientific statements, in particular those that qualify as laws of nature? To decide this question we need to examine the effect of counter-instances on the different kinds of generalisations we have recognised; and compare the actual effect on the beliefs of scientists with the effect of a contradictory instance we would expect if it were assessed according to the rules of classical 'two-valued' logic. The situation according to classical logic can be summed up in the Square of Opposition. Disregarding the complications that ensue if we include existential presuppositions in the table, the Square looks something like this (Table 3.1):

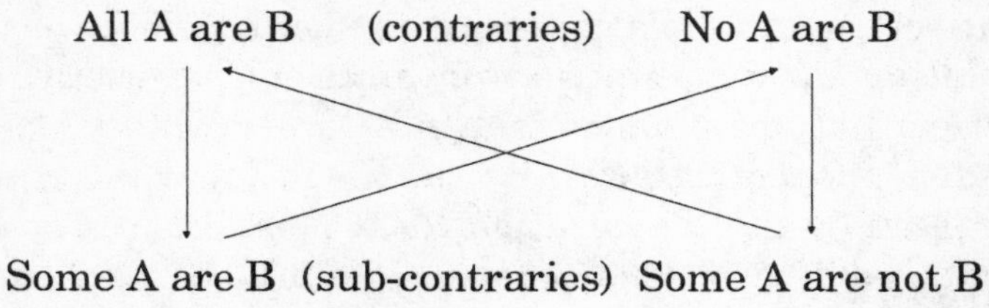

Table 3.1 The classical Square of Opposition.

In this array we have a pair of propositions which are contraries – that is, they cannot be true together, but they can both be false – and two pairs which are contradictories: that is propositions which can neither be true nor false together. If one member of a contradictory pair is true (or false) the other member must be false (or true). According to this system, if we entertain the hypothesis that all *X* are *Y* and we discover that some *X* are not *Y* then we must conclude that all *X* are *Y* is false, and that is all there is to it. This scheme offers no means for distinguishing different effects of the contradiction, which may depend upon which *X* is not *Y*. Let us then look at the effects of contradictory results in scientific research programmes and compare them with this schema.

Counter-instances to taxonomic generalisations

Impressed by what we have seen of birds, we might be inclined to assert the law-like claim that all birds are feathered. Let us take this simple example as a working 'model' of the more technical statements that one would find in a real taxonomy. There are several ways in which the finding of a featherless creature that otherwise seems birdlike would be dealt with:

(1) Having feathers is not one of the defining characteristics of the biological family '*aves*' or birds. A palaeontologist finds a fossil creature that in most respects is birdlike but is apparently covered in scales. In these circumstances it seems to me that we would treat this discovery as disproving the putative law-like claim that all birds are feathered. We might be satisfied that at least most birds were. But the starting point of this pattern of reasoning could be disputed. A dispute might be resolved in several ways:

Either being feathered should continue to be excluded from the differentiae of 'bird'. It should not be used as a characteristic or necessary condition for classifying a creature as a bird. Or being feathered should be included in the differentiae of 'bird'. If a dispute is resolved in this manner we should be obliged to accept that some creatures apparently birds are not true birds. Our original claim can now be revived in modified form as 'All true birds are feathered'.

If the first alternative is chosen we could use 'feathered' and 'not-feathered' as the basis of a new distinction between feathered and featherless birds, just as we use the distinction flighted/flightless to preserve the avian status of the kiwi and the moa.

If the second alternative is chosen we would have to reject the statement that most birds are feathered since it implies, though it does not entail, that some birds are not. Faced with an instance contradictory of our original taxonomic generalisation about some class of entities, we can react to the situation either by rejecting the generalisation as false or by so modifying it that it remains true. Clearly, what move we choose to make depends upon whether we are treating the taxonomic generalisation as potentially analytic or not potentially analytic. If not potentially analytic then we can conclude that the generalisation is false, but if potentially analytic then what has to change is the intension of the class-term 'bird'.

Is the modified hypothesis the same hypothesis or a different hypothesis? In one sense it is the same hypothesis, for it could still be expressed by the use of the sentence 'All birds are feathered'. In another sense it is not the same hypothesis, since the vocable 'bird' now stands for a concept having a different though 'overlapping' sense.

Real discussions like this imaginary one are frequent in science. For instance, in the recent exchanges in Amsterdam between the last-ditch defenders of the hypothesis that HIV is the exclusive cause of the AIDS syndrome and the growing body of sceptics Dr James Curran insisted that unless a sufferer from one or more of the AIDS cluster of diseases tests HIV-positive their disease is not 'true' AIDS. Hence it is analytic that HIV is the sole cause of AIDS. Dr Curran is quoted as saying that a 'true HIV-negative is not supposed to be an AIDS case' (*Sunday Times*, 26 July 1992).

(2) 'Having feathers' is already included among the differentiae of the family *aves*. Our palaeontologist turns up with yet another scaly but birdlike creature. In these circumstances we would surely conclude that this creature is not a bird. Suppose however that, in the quarry from which the first two anomalous specimens came, a great many varieties of featherless birdlike creatures turn up. A dispute might well break out

concerning the starting point of this discussion. At least two possibilities for resolving the dispute are open. Having feathers should not have been included in the differentiae of the family 'birds'. So all the new creatures are assigned to species within the same family as blackbirds, moas and penguins. The other possibility is the setting up of a new family, so that not having feathers is the prime *differentia* of a new family, say *dribs*, or *seva*. The real-life version of this imaginary dispute can be found in the case of *archaeopteryx*, which combines feathers with a jaw full of teeth. Is it a bird or isn't it?

In both the cases in which I have imagined anomalies to have appeared the superficial conclusion was that 'All birds are feathered' is false. But the patterns of reasoning in science allow us to dodge that conclusion. Formal logic recognises only the superficial pattern of reasoning:

> From 'Some S are not P' is true
> we conclude
> 'All S are P' is false.

Clearly formal logic does not have the resources to express all the possibilities that I have discussed. The problem is to decide what is missing from the formal schemata which is present in the actual reasoning. The problem is not hard to solve. What is missing is what, as a matter of fact, the expression 'bird' means to the community who are using it. Scientific reasoning depends for its cogency upon two necessary conditions:

(a) That premises and conclusions of arguments are related according to the rules of formal logic. Let us follow tradition and call this the *validity* requirement.

(b) That the expressions used in those premises and conclusions are related according to the appropriate rules of linguistic usage. Let us call this the *cogency* requirement. To distinguish cogency from validity we need to look at the conditions which are both necessary and sufficient for our accepting an actual piece of reasoning.

Suppose we are sitting on the veranda of the dak bungalow with the District Commissioner. He receives a phone call from one of his assistants upcountry, who gives him the message: 'There is torrential rain in the hills.' The DC takes this message as a premise and concludes: 'There will be floods in the plains.' His conclusion is based on his belief that the conditional statement 'If it rains in the hills there will be floods in the plains'. One of his guests is a logician who construes the reasoning as follows. The DC's reasoning will be valid in three cases: when the premise and the conclusion are both true; when the premise and the conclusion are both false; and when the premise is false and the conclusion true. Russellian formal logic recognises only one case in which the reasoning above is invalid: when the premise is true and the conclusion false. Do these Russellian truth conditions pick out patterns of reasoning all of which we would be ready to accept as cogent? I think not. The third case of a valid pattern would hold if it there had been no rain in the hills but an ancient dam had burst just as the assistant was delivering his erroneous message. The relief efforts organised by the District Commissioner would no doubt have been welcome, but we would hardly grant that they had been soundly based, though, believing in the reliability of his subordinate, he had taken the premise to be true. What is required for the cogency of a pattern of reasoning is all that is involved in saying that there is a causal relation between rain in the hills and floods in the plains. Unfavourable truth-conditions – that is if the premise is true but the conclusion is false – count against the causal relation, but not all logically impeccable truth-conditions count in favour of it.

Counter-instances to non-taxonomic generalisations

It will be remembered that non-taxonomic generalisations are those in which the members of the extensions of the terms that figure in these generalisations are, or could be ordered. It is specially characteristic of the physical sciences that particular observations and experimental results are usually orderable, while in the biological sciences, though classification is an important procedure for these sciences, it is the classes of organisms which are ordered, not the particular organisms which are the extensions of those classes. The physical sciences are largely concerned with the study of those properties of things and substances that vary in degree, such as velocity, temperature, wave-length and so on. Measurement, comparing the current state or degree of a property with a standard or unit, yields quantified, and hence numerical, results. Where there are differences in degree and numerical measures of these differences there is a 'natural' order in that which is measured, roughly from less to more, from smaller to larger, etc. The most common type of law to be found in the physical sciences expresses the mutual variation between variable properties. In the physical sciences we are often concerned with ranges of particular cases. Within certain limits, these ranges extend indefinitely from the cases where the original observations of a mutual variation were carried out. Most laws of nature started life at room temperature, so to say, and have been tested progressively at both higher and higher and lower and lower temperatures. We shall see that the way the result of an experiment or observation affects our belief in a law of nature of this kind depends very much on which part of the range of its variables the experiment was performed. However, it is characteristic of the taxonomic generalisations that we find in classification systems that any member of the extensions of its terms is as good as any other for the exemplification of the type, and so for its suitability as an object of study.

It is characteristic of non-taxonomic generalisations that it does matter from what part of an ordered range of cases an instance is taken. The position of an instance, favourable or unfavourable, in the ordering is an essential ingredient in the pattern of reasoning through the force of the discovery of such a case is assessed.

Counter-instances to putative non-taxonomic laws of nature

Suppose we design an apparatus to determine the law of mutual variation between thermal and electrical conductivity. Provided that we experiment with temperatures around those commonly encountered in the laboratory, we find that the following relation holds:

$\mathbf{T}/k = \mathrm{T} \times \text{constant}$

where **T** is thermal conductivity, *k* electrical conductivity and T the absolute temperature. Notice too that this law, called the Wiedemann-Franz Law, is derivable from Drude's theory of metallic conductivity, a theory constructed by treating the 'free' electrons in a metal as behaving analogously to the molecules of a gas. The extension of the general term T is the ordered range of temperatures from absolute zero.

Experiment discloses that there are the following counter-instances:

(1) At temperatures close to 0°K, $k \rightarrow 0$, and the Wiedemann-Franz relation fails to hold;

(2) At temperatures high with respect to temperature conditions commonly encountered in the laboratory *k* falls off more rapidly than the law predicts.

At each end of the temperature range experiment discloses counter-instances to the law. Suppose we represent the Wiedemann-Franz relation simply as *L*. We have discovered experimentally that there are cases in which not-*L*. According to the rules of classical logic we should represent the situation as follows:

For all temperatures *L*
For some temperatures not-*L*.

We can conclude both

For all temperatures *L* is false (A)

and

Not for all temperatures L is true. (Q)

But neither move, formally valid though each may be, represents the pattern of reasoning we would find in science. (A) does not, because it obscures the fact that we would still wish to claim that for some temperatures L is true. (Q) does not. Since 'Some A are B' and 'Some A are not B' can be true together the unfavourable experimental results do not force us to reject the favourable results that prompted our original hypothesis. But if we express these results in the pattern above we fail to record the most crucial scientific fact: namely, for which temperatures L is true.

This situation arises quite generally with non-taxonomic generalisations. For example, if we consider the effect of counter-instances upon say

Pressure x Velocity/Area of flow = constant

we find, experimentally, that provided the values of the variables fall within certain regions of the ordered ranges of pressures, velocities and areas the relation holds. But should any or all of the variables be given values outside these regions the relation holds no longer. Instances of this kind of thing could be multiplied indefinitely. It is the typical relation between the non-taxonomic generalisation and the values of its variables.

Patterns of reasoning with counter-instances

I began this chapter by pointing out that there are

three dimensions of generality along which the laws of nature are at least initially supposed to apply. They are taken, at first, as applying to all regions of space and time, to all material substances and in all experimental conditions. In this section I shall be considering only the dimension of generality of experimental conditions. In practice we can recognise three different responses by scientists to the discovery of counter-instances to the extension of a law to all possible experimental conditions, and three different kinds of inference-pattern employed consequent upon the discovery of a counter-instance.

1. Limitation.

The counter-instances are used, not to lead us to a rejection of the relation expressed in the law, but to lay down the limits in which it holds. And this is only possible with non-taxonomic generalisations where the ranges of the variables are ordered. Formally we can express limitation thus:

Upper Bound

L under all conditions c_0 ... (Hypothesis)
not-*L* under c_n ... c_r

L under conditions c_0 ... c_{n-1}

Lower Bound

L under conditions c_0 ... c_{n-1}
not-*L* under conditions c_0 ... c_k

L under conditions c_{k+1} ... c_{n-1}

It should be noticed that there is an assumption

involved in the upper-bound cases, not involved in the lower-bound cases. The assumption is that if the relation *L* is found not to hold for some set of cases in the upper regions of an ordered set of conditions, it will not be found to hold in conditions still further removed from those conditions in which we know experimentally that it does hold.

2. *Modification*

Having discovered bounds to the ordered set of conditions in which a relation holds between observables, a new relation is formulated by combining the old relation with the boundary conditions. *A fortiori* the new relation holds for the conditions in which the old relation held, and for some part, as yet undetermined, of the conditions for which it did not hold: for example, the incorporation of the high-pressure and low-temperature boundary conditions on the general gas equation into that equation to formulate Van der Waal's equation; the incorporation of the boundary conditions on Newton's Laws of Motion, that is that these laws must be covariant under the Lorenz transformation into those laws to yield the laws of Special Relativity; the incorporation of the boundary conditions on the laws of the viscous flow of drops, in the form of the effect of internal circulations, to yield Mathur's relation between phase viscosities, drop velocity and the drop deformation constant. From the most commonly known laws to the most obscure we can recognise the modification technique being applied.

Formally we can represent the process as follows:

3. Generality

L under conditions c_0 .... (Hypothesis)
not-L under conditions c_n ... c_r
not-L under conditions c_0 ... c_k

(L & c_0 ... c_k & c_n ... c_r) under conditions c_0 ...

For this law we renew the search for boundary conditions.

3. Idealisation

When we idealise we use the boundary conditions to define, by negation, an ideal universe: ideal with respect to the original relation L. The boundary conditions are interpreted, in the appropriate theory, and a universe is imagined in which this interpretation of the boundary conditions is denied. Ideal gases, perfectly rigid bodies, frictionless pulleys, etc., are arrived at in this way. There can be no logic of idealisation, since this move works, not through some standard pattern of reasoning, but through the construction of an idealised model of the world, a procedure outside the scope of this book, and only to be briefly touched upon in Chapter 4.

The effect of favourable instances on a non-taxonomic law

To find a favourable instance within that part of the ordered range of cases where we already know a functional relationship to hold is useless for science, since it adds nothing to our knowledge, in the dimension of generality with respect to experimental conditions. Of course it may add something to our knowledge of spatio-temporal generality. What would add to our knowledge in this dimension of generality of experimental conditions is a favourable instance in a part of the range of ordered experimental conditions where we have not, as yet, experimented. To accommodate this sort of case we need to add another feature to our schemata for reasoning with counter-instances. We have to distinguish explicitly between known and hypothetical cases. Suppose our putative law to be again L. Suppose that we know, by experiment, that within the range of experimental conditions c_0 ..., L holds for the conditions $c_1 \ldots c_m$. We suppose that no experiments have yet been done in conditions $c_0 \ldots c_{i-1}$, and $c_{m+1} \ldots$ The situation can be expressed as follows:

$$Lc_0 \ldots c_{i-1}(\text{hyp});\ c_i \ldots c_m\ (\text{def});\ c_{m+1} \ldots (\text{hyp}).$$

The effect of a favourable instance in the range $c_i \ldots c_m$ is nil. (Repetition of classical experiments, except for sharpening up the value of numerical constants, is of educational value only.)

The effect of a favourable instance, in either $c_o \ldots c_{i-1}$ or $c_{m+i} \ldots$ is to extend the range of definite cases into the range of hypothetical cases. This can be expressed as follows:

$$\begin{array}{lr} Lc_0 \ldots c_{i-1}\ (\text{hyp});\ c_i \ldots c_m\ (\text{def});\ c_{m+1} \ldots (\text{hyp}) & \\ Lc_k\ (\text{def}) & k<1 \\ Lc_r\ (\text{def}) & >m \\ \hline Lc_0 \ldots c_{k-1}\ (\text{hyp});\ c_k \ldots c_r\ (\text{def});\ c_{r+1} \ldots (\text{hyp}) & \end{array}$$

Favourable instances alter the boundaries between definite and hypothetical applications of a putative law. Counter-instances put boundaries to the range of hypothetical plus definite cases.

Limiting cases of these patterns give us Aristotelian logic. From the hypothesis Lc_0 ... c ... c_m (def); c_{m+1} ... (hyp) we can deduce Lc_i, Lc_{i+1}, ... , Lc_m; any of which could be an S of which it is true that some S are P. Similarly, should we find that a random sample of cases c_0 ... are all cases in which *L* does not hold (and here *L* must be an hypothesis with no evidence at all to support it) we can infer that Lc_0 ... is false. So there is a degenerate case in which from the truth of 'Some S are not P' we can infer the falsity of 'All S is P'.

A summary of our conclusions

Two main results have been arrived at.

(1) The patterns of reasoning appropriate to taxonomic generalisations are correctly represented by the rules of classical logic, as far as the formal relations between particular and general propositions are concerned. But, when we come upon a counter-instance to a generalisation, we can save the putative law, if need be, by treating it as potentially analytic and claiming it as a definition. This pattern of salvation cannot be expressed in terms of the formal relations between the propositions expressing the generalisation and counter-instances, but requires us to consider the meaning of substantive terms, and the consequent intensional relations between generalisation and counter-instances.

(2) The effect of counter or favourable instances on generalisations over ordered sets of conditions is not even formally expressible in the patterns of reasoning

sanctioned by classical logic. The essential true/false relations of that logic are not used.

These conclusions can now be used to discuss the general form of a law of nature with which we began. It was

> For all substances, and for all experimental conditions, and for all space-time locations *L*.

Each quantifier was said to represent a 'dimension of generality'. In practice, laws of nature are often subject to substance-restrictions (only those elements which conduct electricity ...), experimental restrictions (the Wiedemann-Franz Law applies only if the temperature is above 4°K.) and perhaps spatio-temporal restrictions (that the gravitational acceleration is 980 cm/sec/sec in the vicinity of the earth). We studied, in detail, only the restrictions on the dimension of experimental conditions generality.

It should now be easy to see that we can bring together the two facets of this study, that concerning the patterns of reasoning for taxonomic generalisation and those for non-taxonomic. The dimension of substance generality is governed by the rules of reasoning for taxonomic generalisations. Any sample of a substance will do, and in a group of like substances – the halogens, for instance – any member will do to instantiate the generalisations we wish to make about halogens.

What about the dimension of spatio-temporal generality? In so far as there are places and epochs in which certain laws of nature do not apply, such as the first one millionth of a second after the big bang, the patterns of reasoning to accommodate such exceptions are just simplified forms of those I have already laid out for the dimension of generality of experimental conditions. The more remote and the further away the testing grounds for a putative law are, the better it is established. Failure at a distance is accommodated by delineating a boundary at or about that point.

The differences between the patterns of reasoning for each dimension of generality can be further established by considering how generality is saved in each dimension.

Generality in the substance-dimension is taxonomic generality, for by hypothesis, each specimen of a substance is, with respect to the properties of that substance, identical with any other. Hence substance-generality is subject to the logic of taxonomic generalisations, and hence substance-generality can be saved in the face of counter-instances by treating the law as defining or partly defining the substance in question. For instance, if the laws of magnetism fail to hold for phosphorus, we don't drop the laws of magnetism but declare that phosphorus is not a magnetic substance.

Since the conditions under which a law holds are orderable, we can save the law in the face of counter-instances by treating a counter-instance as indicative of the boundaries to the range of experimental conditions for which the law does hold, or might hold. The appropriate logic, then, for this dimension of generality is that of the non-taxonomic generalisation. We can proceed by limitation, modification or idealisation.

Reasoning in the dimension of spatio-temporal generality seems to fit in nicely with the patterns we have discerned in the case of non-taxonomic generalisations. But there is a further feature of spatio-temporal generality that must be acknowledged. This is the role of the Principle of the Uniformity of Nature. This principle takes different forms in different sciences. In cosmology, if phenomena appear to be differently related in different regions of space and time we do not drop the Uniformity of Nature principle, but use it to postulate disturbing influences, such as 'dark matter' or 'massive black holes'. Another strategy that preserves the principle in the face of apparent exceptions is evident in high-energy physics and quantum mechanics. In this branch of physics we declare that the types of individuals and substances with which we are concerned are different in important ways from those with which we are familiar in the world of everyday life. These strategies in this dimension of generality include as an assumptive premise the Principle of the Uniformity of Nature. To question this principle in that dimension would be quite clearly self-defeating.

I conclude then that given the three different dimensions of generality, into which every law of nature is generalised, there can be no one system of logic which will govern inference to or from every law of every kind. Substance-generality requires not only formal but intensional relations; experimental-conditions-generality requires different patterns of reasoning from those employed in the dimension of substance-generality; and spatio-temporal-generality involves an hypothesis not required by either of the other dimensions, in neither of which is the Uniformity of Nature assumed.

4

Modality

The problem of necessity

Incorporated in the very idea of 'law' is some notion of necessity. What laws of nature describe (or in the case of society, prescribe) is somehow obligatory. Whatever the law says *must* happen, hold or obtain, everything else being equal. If this outcome is required by the law then no alternative outcome is possible. This 'necessity' is what is meant by the modality of a law. Nature is not a conscious being to be bound by prescriptions. So it is not at all obvious what the necessity of the laws of nature might be. There are at least two questions for philosophers that the idea of the necessity of laws of nature seem to raise. What do such expressions as 'must' or 'necessitate' mean in this case? If we succeed in making sense of the idea of natural necessity we can then ask what sort of reasons could be given for concluding that a general statement merited the accolade of necessity, that is should be taken as a law?

We can begin by contrasting two interpretations of those statements we have picked out as laws of nature. According to the positivist interpretation they are simply summaries of what has already happened and is going to happen. There is no necessity in nature. The 'must' in a law expresses only the psychological state of those who believe it, a habit of expectation. According to the realist view the laws of nature tell us what must happen, everything else being equal. The 'must' in a law of nature does take its meaning from something in the material world. In their unreflective moments it certainly seems to be the case that scientists, and indeed people generally, take the laws of nature to express real necessities, though they would be hard put to it to say just what these were. The most natural reading of the law of gravity is that two bodies must exert a mutual attraction inversely proportional to the distance between them, and proportional to the product of their masses. A body must fall towards the centre of the earth with such-and-such an acceleration if all supports and hindrances are removed. According to the positivist interpretation of science this commonsense belief is an illusion. The fact that a law of nature is supposed to hold for every case within the bounded dimensions of generality we identified in the last chapter does not pre-empt an answer to the question of the nature of its modality. Something might always be the case without its necessarily being the case. Could the laws of nature be picking out nothing but long-running coincidences? The most famous exponent of the positivist view of the character and meaning of natural necessity is surely David Hume. He tackles the question from the point of view of his analysis of causality.

Hume on causality

According to Hume (1739), the empirical content of a law of nature, what it tells us about the world, is no more than a summary of what is or will be the case. Yet he acknowledges that when we try to express our knowledge of causal processes we tend to express it in necessitarian terms. We do speak of natural necessity, of causal powers and so on. Hume is far from denying that the notion of necessity is part of the *notion* of causality. But he argues that necessity is to be interpreted in a quite different way from that proposed by those who believe that there is a necessity in nature.

Hume analyses the notion of causality into several components. He is concerned almost wholly with causal relations between events. First of all, when we say that A causes B, we have in mind that events like A are usually found to accompany events like B. This is the regularity of sequence component. Furthermore, we seem to require that the cause is, so to say, close to its effect in space and to a lesser extent in time. But the notion of necessity which we take for granted in our ordinary ways of thinking is richer than the conjunction of the two notions, regularity of succession and adjacency of cause to effects. Both these aspects of a causal process can readily be observed in nature. Hume explains the origin of the necessity we all take to be an essential part of the idea of causality, and so of the laws of nature that describe causal relations, as a psychological phenomenon. Regularities impress themselves upon us to such an extent that we come to expect that events will continue to follow the patterns that they have already displayed. Thus we come to expect an event of the same type as the effect on the occurrence of an event of the same type as the cause. It is this expectation that is the real source of our sense that causal processes are necessary, according to Hume. There is no natural necessity. There is only the psychological phenomenon of expectation. In this way Hume accounts for the common-sense must which we quite unthinkingly and naturally tend to incorporate in our statement of natural laws.

What is natural necessity?

Hume shares with all philosophers who have considered the question of the necessity of the laws of nature the realisation that there must be something over and above the data we gather from experience and experiment to account for the necessity we attribute to laws. According to Hume this necessity is not in nature but in ourselves. Those who, like myself, wish to defend the idea that there is genuine natural necessity and that it is a feature of the physical world, are obliged to find some natural property or process, over and above observed regularities of adjacent events, but bound up with them, which embodies that natural necessity. As a natural process occurring, in many instances, independently of human beings, it will not be reducible to some aspect of the psychology of the experimenters.

If there is to be something which we can properly call 'natural necessity' we must distinguish it sharply from logical necessity. Hume has a very powerful argument that counts against any assimilation of the one idea of necessity to the other. The argument is extremely simple. He asks us to consider some sequence of events which we believe to be causally connected: for example, the pouring of acid into an alkali solution, and the subsequent formation of salts. Could we imagine, says Hume, performing the causal or initiating procedure and the expected outcome simply not occurring? Or could we imagine the effect coming about without being preceded by that familiar procedure? In many cases the answer is 'Yes, quite easily'.

We could imagine releasing a stone and its remaining suspended in mid-air, or switching on the current and the galvanometer remaining inert. If that is generally the case, we can conjoin a description of a cause with a negation of the description of its usual effect without producing a contradiction. Or a description of the effect conjoined with the negation of a description of its usual cause. If all necessity were logical necessity this would be impossible. This simple argument shows that the necessity we are looking for is not the necessity of logic. If it is not that kind of necessity that is expressed in the principle of non-contradiction, what is it?

Leibniz on the laws of nature

Leibniz's treatment of the central problems of the philosophy of science is based on the idea of 'possible worlds'. His technique is to set up more and more stringent conditions which possible worlds must meet. These conditions have a theological cast, though they are overtly expressed as formal principles. We can rank the conditions he sets for the acceptability of possible worlds. When we reach the most stringent condition of all only one possible world can meet it, this world.

The weakest condition for eliminating impossible worlds from our imagined repertoire is the principle of non-contradiction. In the *Monadology* (Leibniz, 1714: sections 33-5) Leibniz argues that there are two kinds of truths. Truths of fact must be assessed by reference to some observable state of affairs. Truths of reason can be assessed by analysis and testing against the principle of non-contradiction. In particular among the truths of reason are 'primary principles... [which] have no need of proof and these are identical propositions whose opposite involves an express contradiction'. A huge variety of possible worlds can be imagined to exemplify such 'primary principles' and so to be compatible with the principle of non-contradiction.

The next strongest principle for Leibniz is the principle of sufficient reason. Using this principle we can reduce the multiplicity of possible worlds enormously. As Leibniz says, 'our reasonings are founded on two great principles, that of contradiction and that of sufficient reason, in virtue of which we judge that no fact can be found true or existent, no statement veritable unless there is sufficient reason why it should be so and not otherwise, although these reasons usually cannot be known to us'. Many fewer possible worlds can meet this condition than meet the condition that their description does not contain any express contradiction.

In the correspondence with Clarke (Alexander, 1976) and elsewhere in his many works, Leibniz adds a further principle of yet greater stringency. In considering the question of whether God could have located the universe at some point in space and time other than where and when He did, Leibniz makes use of the principle of the identity of indiscernibles. According to this principle no two physical entities have all their properties in common. Applying this principle reduces the variety of possible worlds still further. For example, Leibniz says (1765: 238), 'it is always necessary that besides the difference of time and place, there should be an internal principle of distinction and though there be several things of the same species, it is none the less true that there be none perfectly similar'. Thus all those possible worlds in which there are a multiplicity of identical entities are ruled out.

And yet this does not finally identify our world. There are many possible worlds compatible with the principle of the identity of indiscernibles. Another, yet stronger principle is needed. It can be expressed in many different ways. Let us call it the 'principle of maximal compossibility'. Leibniz held that, of all possible worlds compatible with the first three principles, only one was compatible with the fourth principle: namely, that in which all those things that could exist compatible with each other do exist. And this finally leads us to our world since he believed that only one world of all possible worlds met this prescription.

It is clearly an implication of Leibniz's arguments that the characteristics we find in this world are there of necessity. Whatever features it has it must have, since we can deduce these features from the four principles successively applied. But what is it that lies behind these principles? Why should we believe that the maximum compatibility principle is true? Leibniz's philosophy of science is intimately linked with his theology. It is God who has sufficient reasons for doing things, and in particular for creating worlds this way or that. It is God whose nature requires that all that is mutually possible should exist. Considered independently of their creator there is nothing to be found in naturally occurring things, substances or processes which imposes this constraint upon possible worlds. Without the hypothesis of a benevolent and logically minded deity Leibniz's case for a necessity in nature seems decidedly thin.

In recent years the Leibnizian way of understanding what a claim of necessity amounts to has been refined by a number of logicians, the most prominent of whom has been Kripke (1963). Logicians had shown that some of our intuitions about the way the concepts of 'necessity' and 'possibility' are used could be captured in the formal rules of logical calculi, in particular as extensions to the system proposed by Russell and Whitehead. By adding a variety of axioms through which rules for the use of the modal concepts could be expressed a number of different concepts of necessity and possibility could be represented. Kripke showed how the differences between the enriched logical calculi could be expressed in the metaphor of Leibniz's 'possible worlds'.

The trick is to ask how wide a swathe of possible worlds is needed to fulfil the requirement that this or that principle of the logic of modality is true in all of them. For example, consider the question: In which possible worlds is it the case that what is necessary is necessarily necessary? A statement, 'p', is necessary if it is true in all possible worlds. So for the statement 'p is necessary' to be itself necessary it too must be true in all possible worlds. Kripke showed that this held in a wholly unrestricted set of possible worlds. By imposing restrictions on the set of possible worlds in which such questions are to be considered, restrictions built up out of the idea of relative possibility, Kripke was able to show that the different logical calculi expressing different modal concepts could be distinguished by reference to the relative breadth of the sets of possible worlds in which they were true. The technical details need not detain us in this book. They are clearly and comprehensively explained in Loux (1979: 21-4).

Suppose we apply the idea of necessity as truth in all of a set of possible worlds to the concept of a law of nature. A statement will be naturally necessary, that is will be a law of nature, if it holds true in a certain set of possible worlds. But how is the scope of that set of worlds to be drawn? It cannot be all possible worlds in the unrestricted sense, since that would fail to distinguish logical truths from laws of nature. It must be something like this: a statement is a law of nature if it is true in all those worlds which differ only as to their initial conditions, that is in which the underlying mechanisms of nature are the same. This restriction amounts to a restriction of the type-hierarchies from which we can create models of the causal mechanisms which lie behind the processes which our laws describe. So whether a statement is a law of nature or not, even in the possible worlds picture, is determined by the type hierarchy principle through which a scientific ontology is controlled. We can also ask: 'Is it true in all these possible worlds that the statement "L" is a law of nature?' Given that the statement 'L' is rooted in a picture (type-hierarchy) common to every one of the worlds in our reference set, the answer must surely be 'Yes'. It follows that for the restricted set of worlds, restricted by the requirement that the natural kinds constitutive of them conform to a certain type-hierarchy, that it is naturally necessary that this or that statement is naturally necessary. We distinguish natural necessity from logical necessity by reference to the way that each set of possible worlds is demarcated.

A 'naturalistic' account of natural necessity

Is there a third way through which a notion of natural necessity can be developed which is neither reducible to psychological states which reflect habits of expectation nor derives from speculations about the nature of God's relation to His creation? To see how this might go, we must return to the first posing of the question as to what we might be getting at when we use 'must' in expressing a law of nature.

If we think about Hume's way of separating what is empirical in experience of the world from what derives from our psychology, we are left with mere sequences of events which, so far as our knowledge of the world goes, could diverge from their usual courses in any way whatever or simply cease to occur at all. Is there any guarantee that we could derive from Nature herself which would support our confidence in the continuity of the regularities we have discerned in the nature and behaviour of physical things and substances into the future and would justify our projection of them back into the distant past? Well, it is no good looking to the uniformity of nature as a general principle which might guarantee such projections of these regularities since that is itself a regularity much in need of guarantees. The same argument that reduces, say, Snell's Law to a regularity between the angle of incidence and angle of refraction in the passage of a ray of light from one medium to another to a mere contingency would 'hit' any suggestion that we can justify our beliefs in the future uniformity of nature from the fact that we have observed many natural processes to repeat themselves so far. This is simply a piece of fallacious reasoning. If we cannot find necessity at one level, the level of what we can observe, where can we find it at another?

To make progress in our search for a way out of this conundrum, we must look more closely at the actual practice of science. Where does our confidence in the necessity of a regularity come from? In discussions of the application of the results of statistical studies to practical life there is a striking clue. It turns out that when people argue about the force of some correlation, say that between smoking and the twenty-odd diseases which have been found to be connected with it statistically, the issue turns almost always on the mechanism by which the inhaling of tobacco smoke induces the cancerous proliferation of cells, the furring of the arteries, the wrinkling of the skin, the development of cataract and so on. In the absence of an understanding of the mechanism or mechanisms involved, it is always open to a critic of the force of the correlation to suggest that the statistics are mere coincidence, or that they reflect the joint effects of a common cause. Nervous people smoke more, and nervous people are more prone to cancer, so smoking is correlated with but not causative of lung diseases. One notices the same sort of arguments in the AIDS controversy. There is a weak correlation between the presence of the HIV virus, as revealed through tests for its characteristic antibodies, and the development of the complex syndrome we have come to call AIDS. What is missing from the story, despite the enormous amount of research put into the matter, is an adequate understanding of the mechanism by which the HIV virus could bring about the AIDS syndrome. Until that mechanism is established, if it ever is, the kind of doubts expressed by Duesenberg and an increasing number of others will remain viable. It is simply disingenuous of the virology establishment to deny their force. What is at issue here is the significance of the scientific 'must'. Why should one assert that those infected with HIV must develop AIDS eventually, *ceteris paribus*?

We can think of the growth of scientific knowledge in a two-dimensional or hierarchical way. First of all, experiment is devoted to establishing the existence of regularities among observed phenomena, either in nature or in the laboratory. But that is never the end of the matter. A second class of experiments is always projected, and undertaken if technically possible, guided by a model of the underlying process which is not generally observable at the time of the formulation of the theory, to try to discover a mechanism through which the cause, identified experimentally and perhaps statistically, produces its effect.

What mechanism would account for the patterns of inheritance described by Mendel's Laws? The 'gene' hypothesis or model, with its distinction between dominant and recessive genes, its identification of the genetic material with unobserved structures in the chromosomes that divide and reconstitute themselves in meiosis and mitosis, if it could be shown to be an accurate portrayal of real natural processes, would provide just that independent backing which would support our confidence in the necessity of the working of Mendel's Laws. Confined to the results of horticulture and stockbreeding, the student of heredity can go no further than Hume's patterns of contingent correlations, and the expectations they induce in those who frequently observe them. How does it come about that a magnetic field induces a current in a moving conductor? How does it come about that 'acid + base = salt + water'? These are the questions which the second phase of a scientific investigation is directed to answering.

What sort of answer do scientists come up with? Well, they come up with descriptions of the mechanisms by which the cause, be it event or state or potent thing or substance, brings about the effect, event or state. These mechanisms usually consist of hypothetical entities which are referred to in the theory that is constructed to account for the experimental results. At the heart of a theory is a model or representation of the relevant mechanism.

We now have the wherewithal for a general account of the modality of scientific laws. Is there a reason over and above the observed data for believing in the necessity of the law, of saying that such and such an event must happen in the appropriate circumstances, everything else being equal? It would be too much to suppose that the extraordinary degree of regularity which has been found to obtain behind even the most apparently disorderly of natural processes is a mere coincidence. That order exists because there is a natural mechanism or, in the case of the 'laws' of a taxonomy an inner nature or constitution, which produces the appearances we observe or evoke with our experimental manipulations. How do we know about such mechanisms? They are described, albeit often in a rather sketchy and improvable way, in the associated theory. The laws of nature and the principles of taxonomies are expressions of necessity.

We can now see the difference between this kind of necessity and logical necessity. Hume was surely right to assert that there is no contradiction in conjoining a description of a cause with the negation of a description of its usual effect. However, it does not follow from that that the law of nature describing that causal relation picks out a mere accident or coincidence. Yet

Leibniz's invocation of the goodness and wisdom of God as a justification for ascribing necessity to the laws of nature is not the kind of consideration which is popular at the present day. The third way offers an account of natural necessity that is neither psychological nor theological. The more comprehensive and powerful the theory, the more we are justified in the use of the scientific 'must'. But if this account of natural necessity is to be convincing much more must be said about theories.

What is a theory?

In the natural sciences there are two very different kinds of theory. There are formal, mathematical structures involving algebraic representations of systems of laws. For example, elementary kinematics is just a theory. From an assumption about acceleration, say that it is uniform, we can derive a bouquet of kinematic laws by applying the rules of the integral calculus.

$$d^2s/dt^2 = k$$
$$ds/dt = kt + c$$
$$s = kt^2/2 + ct + q$$

These three laws form a deductive chain. We could call theories of this sort 'reticular', since they serve to link together the concepts of the theory into a kind of network. Such theories say nothing about the causes of the regularities they describe. They have the kind of necessity that we came across in discussing taxonomic generalisations (and the proposals of Poincaré). The strategy for managing apparent exceptions is to invoke another physical process which disturbs the smooth running of the process described by these laws. For instance, air resistance could be invoked to account for observed deviations from the third of the above laws, when it is applied to the motion of a body falling near the surface of the earth. We would then add Bernoulli's law to our kinematics.

Then there are explanatory theories, the kind of theories the role of which I have been discussing in this chapter. They are based on models or analogues of the entities and processes which serve as representations of the mechanisms, usually unobserved, which generate the phenomena described in the laws they support. These models are not haphazard inventions. They are subject to stringent constraints on their physical plausibility. To achieve that the concepts or ideas of the types of the entities, properties and processes of the model are usually derived from those existing type-hierarchies which represent the natural kinds of which the scientific community believes the whole world has been constituted. A central question for philosophy of science is the question of the status of such theories: are they to be taken as more or less verisimilitudinous depictions of real processes? That is, do their models represent reality? The idea that they are to be taken as depictions of something as real as the processes they are invoked to explain has been called the thesis of *scientific realism.* In the form in which I wish to defend this idea and make use of it in this book, I shall express the relationship between theories and reality in terms of the concept of verisimilitude. The verisimilitude of a theory is the degree to which the model a theory describes resembles the reality it represents. Without a defence of scientific realism my account of natural necessity in terms of the backing that a theory gives to a description of an observed regularity is hollow.

The latest argument for realism

The defence of a realist reading of science against the massed ranks of sceptics involves many complex and subtle considerations (Hacking, 1987; Harré, 1987; Devitt, 1984; Aronson, 1991; Aronson, Harré and Way, 1993 etc.). I shall simply sketch the latest arguments for realism, laying out just those main points that are germane to the question of natural necessity.

The question of what the concepts of verisimilitude of models, the plausibility of theories and the truth of scientific laws *mean* must be distinguished from the question of *how we know* that one theory is nearer the truth than another, one model more verisimilitudinous than another and so on. The former or semantic questions can be answered independently of and prior to attempts to answer the latter or epistemic questions.

The content of a theory consists of a model together with a type-hierarchy in which the model is embedded. For example the backing for Snell's Law offered by Descartes consisted of a mechanical model through which the kind of the underlying process which explained refraction of light was fitted into a type-hierarchy as follows:

Supertype	Mechanical processes
Subtypes	(1) Impact of bodies (2) Other actions
Subtypes of (1)	[(a) Balls/racquets (b) light corpuscles etc.]
Subtypes of (2)	[(a) Friction (b) Mixing of fluids etc.]

Through his analogy with tennis Descartes contrives that his model is a subtype of the supertype 'mechanical process'.

A theory is plausible if it is empirically adequate: that is, facilitates successful predictions and retrodictions. But it must also be metaphysically adequate: that is, the model that is used to represent unobserved processes must be embedded in a hierarchy of established natural kinds. Empirical adequacy alone is not enough. For any database there are indefinitely many alternative empirically adequate theories from which it can be deduced. Only by attending to the question of their relative physical plausibility can a choice be made between them that is scientifically acceptable.

But at this point we encounter the problem that tempts some philosophers to retreat to some form of positivism: that is, to be sceptical about the possibility of judging theories on their degree of plausibility. If we need models to represent those aspects of the world we have not yet or perhaps never will observe, and the degree of plausibility of a theory depends on the degree of verisimilitude of its model, how can we make judgments of plausibility unless we can compare a model and what it represents in the world? But we need the model just because we cannot observe the relevant aspect of the world. So it seems that we cannot make the comparison that we need.

There is an answer through which a general realist attitude to science can be defended but it is complex. When do we require models? The need for them arises in three different circumstances.

(i) We have not yet had an opportunity to make a

comparison between the model and what it is meant to represent. But there are cases in which it is possible in fact as well as in principle to make that comparison. A car mechanic can propose a theory to account for the faulty running of an engine, and his 'model' of the faulty part can be checked by dismantling the motor. This sort of case can be found in geology. For example, an explanation of an observable topography is found by the construction of a model of the structure of the underlying but unobservable strata, which can be checked when the region is mined and quarried. These illustrate the correctness of a realist interpretation of science for the case in which the aspect of the world that is modelled is perceptible in the same manner as the model.

(ii) We have not yet had the opportunity to make the comparison because of the technical difficulties of bringing the processes or structures modelled in our theory to the test of observation. In many interesting cases that bringing to light is possible in principle, if only the technical difficulties can be solved. In such cases – for instance, the early stages of the microbial theory of infections – the natural kind of the world-aspect modelled (bacteria) is of the same ontological kind as the model (microbes). Microbes and cows are both organisms, the one micro (and imaginary) and the other macro (and observable). They are subtypes of the same supertype, and both have a place in a biological type-hierarchy. The test of the model came about when the technical difficulties of observing that which microbes represented were overcome. The imaginary microbes of the theory were pretty good, though not perfect analogues of the bacteria that were visible under the microscopes of Pasteur and Koch.

The history of science demonstrates that the realist approach to science was the proper one to adopt for a vast class of theories the content of which is tied in with the natural kinds that fill out our type-hierarchies. In fact increasing plausibility, in both its aspects – namely, empirical and metaphysical – leads to increasing verisimilitude.

From a philosophical point of view it is worth noting that the argument I have just presented is an inductive argument of a very simple and familiar kind. The more cases that favour a hypothesis that we can assemble, and the fewer there are of any significance that count against it, the more we are entitled to accept it. If inductive arguments are acceptable in natural science (and where would it be without them?) surely they are acceptable in the philosophy of science.

(iii) But there remain many examples of theories the models of which are so constructed that we will never have the opportunity to make the needful comparisons between them and what they represent. In such cases the type-hierarchy upon which the ontology of that branch of science is based and which encompasses both its model and the aspects of the world that they model does not include any perceptible kinds. We have no idea what it would be like to perceive a subatomic particle, a quantum state or a field potential.

However we do have a very good idea of the dispositions and powers of these mysterious entities. Numerous experiments have revealed the affordances of a great variety of apparatus/world ensembles. The type-hierarchies of natural kinds that dominate subatomic and high energy physics are built up in terms of affordances.

We can now take a second inductive step, with the help of one more crucial concept. Let us return for a moment to reflect on the argument that underlies our confidence in the realist reading of theories that are based on models of kinds of beings. Long ago (1660) Robert Boyle pointed out that it is possible to manipulate experimentally instances of those classes of entity that we are unable to perceive. He argued that chemical operations are such that we can justly conclude that when we perform them we are manipulating imperceptible particles, the corpuscles that play the role of atoms in chemistry. In the first two of our three cases above it is evident that manipulation precedes perception. Indeed successful manipulation of the imperceptible aspects of nature is just what the empirical adequacy of a theory shows we have achieved. Taken together, the history of the first two cases of theory-guided research supports the general principle that manipulative success gives a good ground for believing in the existence of those classes of beings we think our experimental operations have manipulated. Technical developments have made it possible to check this principle by observation. Most of experimental physics nowadays involves manipulating beings which could never be perceived, no matter what technical advances are made. We can manipulate magnetic and electric fields, and with them we can manipulate all sorts of subatomic particles to display their affordances in an apparatus that responds with observable changes of its state. The electric field 'pulls' electrons into certain paths which are revealed to an observer in a cloud chamber. The cloud chamber/world ensemble affords tracks. Examples of this sort can be multiplied indefinitely. Every one strengthens the inductive argument that supports the principle that manipulative success together with the ontological plausibility of the type-hierarchy involved gives us good grounds for believing in the existence of the entities we believe we are manipulating. From the point of view of the philosophy of science that is the final step in the argument for realist readings of successful explanatory theories.

Finally, by extending the compass of the notion of plausibility to include the third kind of physical theory just discussed we can justly claim that, in general, the greater the plausibility of all three classes of explanatory theories the greater their verisimilitude.

Summary: applying this account of natural necessity to our four examples

The most natural way to conceive of natural necessity, the 'must' of the laws of nature, is in terms of two kinds of reasons for believing that a statement describing an observed regularity is a law: namely, experimental support and theoretical explanation in terms of a plausible theory. Snell's Law is that much the more necessary if it is further grounded in a theory such as that of Descartes. If a theory is based on a representation of the hidden or occult mechanisms by which the phenomena in question are produced, and we can take that theory realistically, then we do have a serious reason for believing that *ceteris paribus*, if the conditions are satisfied, observable events *must* occur as the law says they will. In the Cartesian case, we have two models, one concrete, the ball and the sheet, which represents the behaviour of light up to a certain degree, and another, derived from it by abstraction, in which we have a yet closer match to what we take the process of light transmission to be. In the absence of theoretical support we are hardly justified in picking out a description of one natural regularity as a law from any other. This idea, that natural necessity is the result of embedding a general statement in a theory, could be called the double level or dual support account of necessity.

Level One: experimentally induced phenomenon.
Level Two: theoretically imagined generative process.

The job of the theoretician is to tie the model inherent in the theory to the induced phenomena by demonstration. We have seen this in the case of Descartes's geometrical demonstration of Snell's Law. For Mendel and Boyle, it was a promissory note until Maxwell tamed the molecules, theoretically, and Watson and Crick uncovered the structure of the genetic material and its processes of redistribution in their analysis of the chemical structure of DNA.

The 'must' of Newton's First Law is very different. There is no deeper level relative to the processes that law describes in which to seek a mechanism which generates uniform motion. There is no such mechanism. That is part of the point of the law. Newton's Law reaches deep, to the very foundations of physics. It gets its necessity from the fact that it specifies what it is for something to be a material thing. It is more like a taxonomic generalisation than it is like a causal law. Of course every body must continue in its state of rest or inertial motion. Anything which does not is thereby struck from the role of materialship. Newton's Law defines or specifies a universe. It is regulative and constitutive, part of the way we specify our world. Unlike our other three examples Newton's Third Law can find a place only in a reticular theory.

5

Puzzles and Paradoxes

Introduction: laws and natural kind type-hierarchies

The animating principle behind the material and discursive practices of science, behind experimenting and theorising, is the thesis that nature exemplifies multiple hierarchies of natural kinds. All experiments on, and all reasoning about natural phenomena, as they are presented to us through the multi-coloured lenses of apparatus and the systematising power of technical vocabularies, depend on this assumption. Its importance becomes even clearer when we examine the famous paradoxes of induction, the paradox of the emeralds and the paradox of the ravens. Both paradoxes concern the relationship between experimental and observational data and the laws they purport to support. In introducing this topic we shall return to our starting point once again with the observation that a law is, among other things, a general statement, descriptive of some seeming natural regularity. The data we have accumulated in support of our belief in this or that law can be expressed as an open set of singular statements describing the observed behaviour of particular things (including apparatus) and the repeated concomitance of types of events. The paradoxes of induction have been discussed with the use of simplified examples, not real laws of nature. The paradoxes infect all general or law-like statements about alleged matters of fact, not only those we have picked out as laws of nature.

The paradox of the emeralds

The paradox can be expressed in the following way (Goodman, 1954): On the basis of a study of emeralds in which it has been found that without exception they are green, we seem to be justified in claiming that all emeralds wherever and whenever they exist are green. But suppose we introduce a new property, 'grue' as follows: Grue emeralds when observed are always green, but those which are unobserved at any time are blue. So, though there are blue emeralds, we shall never actually observe one, however far we push our enquiries. Our long-term observations of green emeralds seem to support two incompatible law-like statements, that all emeralds (including of course those not yet observed) are green and that all emeralds are grue. Goodman introduced 'grue' merely as a logical construction out of the perceptible property concepts 'green' and 'blue'. But if it is to be studied as a possible addition to a scientific vocabulary it cannot itself be a perceptible property for us. Whenever an observation of a grue emerald is made we must suppose that it changes from blue to green. Green emeralds do not change when observed. Neither do really blue sapphires. In introducing the predicate 'grue' we also introduce an additional causal hypothesis into our chemistry and physics; namely, that when observed grue emeralds change from blue to green.

There seem to be two cases. In one case the change

from blue to green, when it occurs, is circumstantial. A rose can change from looking red to looking black just by being removed from ordinary illumination to the light of a sodium lamp. The real essences of the material beings involved remain unchanged. A sheet of white paper which looks green under green light will look white under white light. Suppose we always observe paper under green light, for some reason. In that case there is no inductive puzzle involved since no one who knew the facts of the case would be tempted to say the paper itself was grite, and no reasonable person would be tempted to induce a law-like statement one way or the other from noticing that a piece of white paper always looks green, whenever it is observed. In the other case suppose grue is an essential property, something in the nature of a certain kind of gemstone. In traditional terminology it must be a secondary property, a power to induce a certain phenomenal experience. Powers of this sort are grounded in some feature of the constitution of the being in question. Now Goodman's puzzle does arise, since the current green appearance of a thing is as good a reason to suppose it is really green, that is has a nature such that it permanently possesses the disposition to appear green, as to suppose that it is really grue, that is has a constitutional property such that though it always looks green to us, it has a constitution such that it does have the power to look blue, though to which observers is not specified. As far as people are concerned, this power could only be specified counterfactually. Were people, *per impossible*, to be able to perceive emeralds in ways other than those in which they do observe them they would look blue. To make some sense of this counterfactual, let us suppose that bees (which are sensitive only to the blue end of the spectrum) are brought into the fable. Bees will always see grue emeralds as blue. Interspecific communication on this matter would lead to an impasse rather than a solution to the problem, if, in the positivist tradition, the only admissible properties were observables. People would insist that all emeralds were green, and bees would be equally adamant that they were blue. Each would accuse the other of colour blindness.

Creatures who were capable of perceiving a grue thing as grue would not see it as either green or blue, but as whatever primary quality it was that grounded the disposition to look one way to people and another way to bees. For human beings the difference between green emeralds and grue emeralds could not be settled by inductive evidence based on appearances, even with the help of the bees. Within the framework of the positivist account of science Goodman's puzzle is indeed a paradox. To resolve the paradox it would be necessary, so this fictional tale illustrates, to make reference to unobservable properties defined in a theoretical context. It would be necessary to adopt a realist stance to the interpretation of science. The paradoxical character of Goodman's puzzle in the positivist framework is a *reductio ad absurdum* of that framework. Once we have to hand the distinction between powers, natures and appearances, the common stock-in-trade of scientific thinking in the realist framework, the grue problem illustrates the enormous importance that the concept of a natural-kind plays in real science. Emeralds, as a natural-kind, cannot change spontaneously from blue to green at some arbitrary moment when someone chooses to examine them. This is because as a natural-kind they are defined in the dual contexts of theory and practice as those beings with such a molecular constitution that in normal circumstances they will look green. Martians, as it happens, can perceive the property that we, as theoreticians, have introduced to explain the apparent spontaneous change of colour that occurs when certain very rare gemstones are observed, namely 'grue'. We use the reactions of bees to distinguish them from genuine emeralds. Their philosophers have pointed out that, on the basis of evidence culled from appearances alone, namely that they always look grue, it is just as rational to infer that when unobserved these stones are bleen as that they are grue. For people a gemstone which changed from bleen to grue on observation *would have* looked green all along. Just as grue would be an unobservable primary quality for people, so green would be an unobservable primary quality for Martians.

The green/grue puzzle does not constitute a problem for a realist philosophy of science because the theoretical context of any research programme depends on some scheme of interlocking type-hierarchies. Such a scheme is elaborated until as many as possible of the observed empirical irregularities and discontinuities in some field of interest are accounted for in terms of theoretically defined stable natures. We humans do not believe emeralds will always be green and have always been so because they always have been observed to be green, but because what it is to be an emerald is to have a certain molecular constitution. Things with this constitution usually look green to well-placed people. As Eddington remarked, one should accept only those facts which are well confirmed by theory. We can make claims about natural necessities. That is we can sensibly use modal qualifiers like 'must', 'would' and 'should', to the degree that we are convinced that our theoretical sketches of the natures of things are correct.

The paradox of the black ravens

The best known account of this famous paradox is that of Hempel (1965). Suppose we express our belief in the law-like character of the relation between being a raven and being black as a conditional statement. It would run like this:

If anything is a raven then it is black.

According to the rules of traditional logic this statement can be contraposed to give the following logically equivalent formulation:

If anything is non-black then it is a non-raven.

By one statement being logically equivalent to another it is meant that they match perfectly for truth and falsehood. Contraposition preserves truth value. If the former is true so is the latter, and if the one is false so is the other.

It is important to realise that without further qualification the quantifier 'anything' ranges over an unrestricted universe of entities. What would we ordinarily expect the 'law of ravens' to cover? Should it, for example, be taken to apply to white shoes? Our intuitions tell us that the law of ravens has nothing whatever to do with shoes. At most it says something about birds. Why drag in shoes? Well, if we are permitted to perform the operation of contraposition on the unqualified law statement, whose quantifier ranges over all the universe, then everything in the universe except ravens falls under the category of non-ravens, including, of course, shoes. So if the former is a true statement about ravens the latter must be a true statement about, *inter alia*, shoes.

What should a scientist set out to observe if he or she wanted to find support for the raven law? According to the criterion proposed by Nicod, anything which instantiates the subject term of the law, in this case ravens. The discovery of a black raven is surely support for the law 'If anything is a raven then it is black'. This is called 'the principle of instance confirmation'. The discovery of a white raven would disconfirm the law, unless some accident could be invoked to explain it. Laws are confirmed and disconfirmed by the discovery of instances or counter-instances of the things they purport to describe. So long as we accept both the admissibility of contraposition, the universality of the quantifier and the principle of instance confirmation, we must admit that a white shoe would confirm, to some unspecified degree, the claim that everything which is non-black (e.g. white) is a non-raven (e.g. a shoe). But this last statement is the contrapositive of the law of ravens, and hence is logically equivalent to it. So we seem to be forced to admit that white shoes confirm a law about the blackness of ravens. But this conclusion clashes with our intuition that shoes, irrespective of their colour, have nothing to do with the credibility of law-like statements about birds.

There are four elements in this puzzle: the logical principle of contraposition, the unrestricted scope of the quantifier 'anything', the principle of instance confirmation, and our intuitions about what is relevant to what. It seems that to avoid a contradiction something must 'give'.

The sources of the puzzle seem terribly obvious. Why should we accept that the contrapositive of a law of nature is also a law? And why should we accept that a law of nature purporting to be about a species of bird actually comprehends everything in the universe? Aronson has tried the raven paradox on working scientists, and one and all they reject the idea that the law of ravens applies to the complement class, non-ravens. Something seems to be awry with the way that traditional logic is used in the analysis of law-like statements and in the logical operations with them that it seems to permit.

No entirely satisfactory solution has emerged within the framework of logicism. Can we step outside that framework to find another way of displaying the formal syntax of law statements? Perhaps adopting a different formal apparatus will reveal some other rather different structure, a structure which will not lead to this and other paradoxes that beset attempts at representing the assessment of laws of nature in the apparatus of traditional logic.

Let us try out the idea that quantifiers, as they are understood in traditional formal logic – that is as genuinely universal in scope – are not the best way of expressing the ontological commitments of a statement, despite their canonisation in this role by W.V. Quine. From the point of view of the philosophy of science Strawson's presupposition relation fares much better (Strawson, 1954) in accounting for the intuitions of working scientists. The presupposition relation ties together assumptions about what we take to exist with the very possibility of assessing the statement which presupposes it as either true or false, or in our context plausible or implausible. Thus 'All ravens are black' is either a plausible or an implausible law-like assertion about ravens if there are ravens, but it cannot be assessed for plausibility at all if there are no ravens. But, the critic might reply, surely there is a quantifier buried in this reasoning, for we mean all ravens, not just a few or a lot, but all. True, but this is not a universal quantifier in the strong sense, since it quantifies only over ravens. That restriction is ensured by the incorporation of the ontological clause 'and there are ravens' in the discourse. Presupposition is not a truth-functional relation, since the truth of a presupposition entails only the 'truth or falsity' of what presupposes it. The falsity of an existential presupposition entails that the statement which presupposes it is neither true nor false.

The move to presupposition instead of quantification for fixing the scope of a law-like statement provides a principled way of tying the procedures for finding support for laws of nature to an examination of instances of the natural kinds presupposed by the putative law. It is because ravens are birds that it makes sense to contemplate the possibility of a lawful relation between their species and their colour, and even a naturally necessary relation. The mediator is the theory of natural selection. Another context where kinds and colours might be thought to be lawfully tied to one another is provided by the physical chemistry of gemstones. But despite the fact that green emeralds are undoubtedly correctly described as non-black non-ravens they have no role to play in assessing the plausibility of 'All ravens are black' because their existence is not among the existential presuppositions of that statement. The theory of natural selection does not apply to them, and so has no role to play in backing up the claim that the kind-to-colour relation is lawful for birds. Aronson has developed a more or less identical argument, where he prefixes an ontological statement to the expression of the substantive content of the law. He would not express it in the form of the presupposition relation between existential commitments and the scope of relevant test objects. He refers to this essential background to experimental and observational science as the common ontology of the theory which has something to say about the colours of the plumage of birds. But this is a difference without a distinction. The power of this style of analysis shows up in the way that the generalising function and the ontological function of discourse, elided in the traditional quantifier, are located in separate, functionally and formally distinct elements in the display of the formal syntax of law statements.

There remains the question of the propriety of the expression of the substantive content of the law statement as a conditional statement, our very first step, the step which invited the use of the rule of contraposition. The next phase in this argument I owe wholly to Aronson. Take a simpler example than that of the colour of ravens, say Newton's Second Law, $F=ma$ (force is the product of mass by acceleration). Following the traditional way of formulating a law statement, one would try to express it as

If anything is subject to a constant Force then it

> will accelerate at a rate proportional to the quotient of the Force exerted upon it and its Mass.

In this form the law can be contraposed.

> If anything is a non-accelerating-at-a-rate-proportional-to-the-quotient-of-the-Force-exerted-on-it-and-its-Mass thing then it has not been acted on by a constant Force.

If we place no restrictions on the kind under which we are considering the entities to which this law applies we must admit that it includes not only stationary things and uniformly moving things, but singing things, glowing things, and so on. Finding a glowing thing which is not being acted on by an impressed force, though common sense suggests that it is irrelevant to the assessment of a law of mechanics, should support Newton's Second Law according to the paradox.

However, there is an alternative way of formally presenting the logical syntax of the Newtonian Second Law. It can be set out as a function: that is, as a mapping from one set to another. These sets could be values of variables representing phenomena, or, perhaps better, one could treat a law as a mapping relation between the states themselves. Whichever way the law is taken, $F{=}ma$ is not a molecular proposition expressing a conditional relation between atomic propositions. It is the succinct expression of a mapping relation between sets. Aronson has pointed out that as a mapping relation it has no contrapositive. The case of a body which is not accelerating and is not acted on by a force is already included in the law as the mapping from the first member of the '*F*' set onto the first member of the '*a*' set. Thus 'non-accelerating' means '*a* = 0' – that is, either stationary or moving uniformly relative to the relevant inertial frame – and so cannot mean, *inter alia*, 'glowing'. As a mapping from one set to another the law applies only to those beings which are accelerating because acted on by forces, and tells how they perform. That the law is about them and no other beings is determined by its existential presuppositions.

We now have an alternative formal analysis of the logical syntax of a law statement. There is no universal quantifier prefixed to the sentence form, but instead there is an ontological clause standing in a presupposition relation to the rest of the law statement. The core of the law statement is not a conditional statement, but a mapping from one set to another. This analysis displays a formal structure within which the raven paradox cannot be generated. The ontological clause ensures that any quantifiers are restricted in range to relevant entities, while the mapping relation precludes the operation of contraposition.

It should now be clear that the raven paradox is an artefact of the choice of formal apparatus for expressing the structure and content of the law of ravens. The exposition of a formal apparatus, alternative to the first-order predicate calculus, shows that traditional logic is not the only possible formal syntax for displaying the structure of a law statement. We can also now see why laws of nature can be assessed only relative to the type-hierarchies in which they are embedded. It is only by virtue of the relevant type-hierarchy that the content of ontological presuppositions can be filled out, and the range of beings relative to which the assessment of the plausibility of the statement of a law can be made.

Conclusion

In both cases the resolution of the apparent paradox in the way that evidence bears upon our right to accept a law of nature was resolved in the same way. Laws of nature are always understood as embedded in a background system of kinds, an interlinked ontology of type-hierarchies. In the last analysis we must reject any interpretation of such laws except that in which we read them as singular statements about kinds or types. A law of nature tells us what kinds of things, events and properties must always (all else being equal) go along with what. We have come to see that the 'must' of natural necessity has its place in a law because the law is not an independent, free-standing statement, but is intimately bound up with a theory, in the heart of which is a model or analogy which represents the generative mechanisms that produce the phenomena the law describes.

*

Note: My presentation of the 'grue' puzzle is a simplified version of Goodman's original formulation of the paradox. In his version grue emeralds were green up to some time t and blue thereafter. So, if we were willing to wait till time $t + 1$ we could tell whether we had been right to take the greenness of the emeralds we had examined before t as evidence for the hypothesis that all emeralds are green rather than blue.

References

Alexander, H.G. 1976 (ed) *The Leibniz-Clarke correspondence* Manchester: Manchester University Press

Aronson, J.L. 1991 'Realism, anti-foundationalism and the enthusiasm for natural kinds' *Philosophical studies* **161**

Aronson, J.L. 1992 'Verisimilitude and type-hierarchies' *Philosophical topics*, forthcoming

Aronson, J.L., Harré, R. and Way, E. 1993 *Realism rescued* London: Duckworth

Berkeley, G. 1710 *Treatise concerning the principles of human knowledge* London

Bohr, N. 1958 *Atomic physics and human knowledge* New York: Wiley

Boyle, R. 1660 *New experiments, physico-mechanicall, touching on the spring of the air and its effects* London

Cartwright, N. 1989 *Nature's capacities and their measurement* Oxford: Clarendon Press

Descartes, R. 1637 *La Dioptrique*

Devitt, M. 1984 *Realism and truth* Oxford: Blackwell

Dretsky, F. I. 1977 'Laws of nature' *Philosophy of science* **44** 248-68.

Galileo, G. 1632 (1953) *Dialogue concerning the two chief world systems* trans S. Drake, Berkeley: University of California Press

Galileo, G. 1638 (1914) *Dialogues concerning two new sciences* trans H. Crew and A. de Salvio, New York: Dover

Gibson, J.J. 1979 *The ecological approach to visual perception* Boston: Howard Mifflin

Gooding, D. 1990 *Experiment and the making of meaning* Dordrecht: Kluwer

Goodman, N. 1954 *Fact, fiction and forecast* London: Athlone Press

Hacking, I. 1987 *Representing and intervening* Cambridge: Cambridge University Press

Harré, R. 1987 *Varieties of realism* Oxford: Blackwell

Hempel, C.G. 1965 *Aspects of scientific explanation and other essays in the philosophy of science* New York: Free Press

Hume, D. 1739 *A treatise of human nature* London

Kripke, S. 1963 'Semantical considerations on modal logic' *Acta Filosofica Fennica* **16** 83-94

Leibniz, G. 1714 (1951) *The monadology* trans M. Morris, London: Dent

Leibniz, G. 1765 (1981) *New essays on human understanding* trans P. Remnant and J. Bennett, Cambridge: Cambridge University Press

Loux, M.J. 1979 *The actual and the possible* Ithaca, N.Y.: Cornell University Press

Mach, E. 1894 (1898) *Popular scientific lectures* trans T.J. McCormack, Chicago: Open Court

Newton, I. 1756 (1958) 'The first letter to Bentley' in I.B. Cohen (ed) *Isaac Newton's papers and letters on natural philosophy* Cambridge: Cambridge University Press, p. 280

Newton, I. 1687 (1947) *Philosophiae naturalis principia mathematica* London: The Royal Society; trans A. Motte, F. Cajori (ed) Berkeley: University of California Press

Poincaré, H. 1958 *The value of science* trans G.B. Halstead, London: Constable, Part III

Popper, K.R. 1959 *The logic of scientific discovery* London: Hutchinson

Root Bernstein, R.S. 1983 'Mendel and methodology' *History of science* **21** 275-95

Sabra A.I. 1981 *Theories of light from Descartes to Newton* Cambridge: Cambridge University Press

References

Strawson, P.F. 1954 *Introduction to logical theory* London: Methuen
Tooley, M. 1987 *Causation* Oxford: Clarendon Press
Way, E.C. 1991 *Knowledge representation and metaphor* Dordrecht: Kluwer

Index

also available in the Interpretations series

FACTS

Bede Rundle

Fellow of Trinity College, Oxford

Facts have many contrasts: with fiction and fantasy, with speculation and opinion, with hypothesis and theory. Two contrasts of particular interest – to philosophy, to morality, to the natural and social sciences – are those of fact with theory and fact with value, and a central concern of this book is to examine arguments which would have us enlarge these domains at the expense of the realm of facts.

It is naturally supposed either that facts enjoy a concrete existence, being numbered among such worldly items as events, or that they possess a more abstract character, having their home in language. Both alternatives are the source of difficulties, encouraging sceptical claims that the existence of facts is unknowable and their nature impenetrable. An analysis of *fact* is offered which, in rejecting both possibilities, seeks to remove the mystery which surrounds this key concept.

ISBN 0 7156 2467 9

also available in the Interpretations series

PARADOXES

Justin Leiber

Professor of Philosophy at the University of Houston, Texas

Paradoxes are many things. Artificial intelligence views them as viruses of the brain, strange replicators that unexpectedly exploit design possibilities. For the child, they are intellectual cartwheels, an everyday delight. For mathematicians and logicians, they reveal skeletons in the closet of reason. For philosophers and dramatists, they capture the contradictions of experience. The historian of ideas sees that they come in successive waves, surging through Classical Greece, the Renaissance and the twentieth century.

Professor Leiber's user-friendly guide to paradoxes provides an up-to-date survey of an ancient and perennial source of puzzlement.

ISBN 0 7156 2426 1

also available from Duckworth

MIND MATTERS

Series editor: Judith Hughes

Many philosophy books claim to be written for the general reader as well as the academic, but all too few really cater for their needs. This series explores philosophical issues and is written specifically for the general reader.

Each book starts with the kind of question we may ask ourselves without, perhaps, realising that we are 'philosophising'. Do computers have minds? Can a pile of bricks be a work of art? Should we hold pathological killers responsible for their crimes? These questions are explored and new questions raised with frequent reference to the views of the major philosophers.

The books in the 'Mind Matters' series are concise, lively, inexpensive, jargon-free and, above all, a fascinating read.

ART OR BUNK?

Ian Ground

For many people, the infamous pile of bricks in the Tate Gallery, Carl André's *Equivalent VIII*, is still the most potent symbol of modern art, and it is what inspires Ian Ground's question 'Art or Bunk?' Using his wide knowledge of classical aesthetics and current ideas in the philosophy of art, he guides us through various attempts to say just what sort of thing a work of art is, and shows us a way to answer the question. Although his examples relate to the visual arts, his arguments are applicable to arts in general, and his book provides a much-needed lively and readable introduction to aesthetics.

Hbk ISBN 1-85399-014-0 £17.95
Pbk ISBN 1-85399-015-9 £6.95

BEFORE EUREKA

the presocratics and their science

Robin Waterfield

What kind of science and scientific thinking went on in the West before the familiar names of Aristotle and Archimedes claimed centre stage? Actually an extraordinary amount – extraordinary both in volume and in breadth of scale. Their enterprise was no less than to describe and explain the whole universe and all its major constituent parts. Their achievement was to establish the authenticity of the rational human mind as a tool for tackling the universe in a matter-of-fact manner.

Robin Waterfield not only describes the Presocratic enterprise in a clear and lively way, he also questions what science is, and whether the Presocratics were scientists in the modern sense of the word. As his previous publications range from the sublimity of Greek philosophy to the fantasy of children's space-age fiction, he is well qualified to write this book.

Hbk ISBN 1-85399-074-4 £17.95
Pbk ISBN 1-85399-075-2 £6.95

Mind Matters series

CAN'T WE MAKE MORAL JUDGEMENTS?

Mary Midgley

How many times do we hear the statement, 'It's not for me to judge'? It conveys one of the most popular ideas of our time: that to make judgements of others is essentially wrong. But doesn't this idea itself involve a moral judgement? What is it? Could we possibly avoid making it? Why have so many thinkers urged us to do this impossible thing?

In this lively and approachable discussion, Mary Midgley turns a spotlight on the fashionable view that we no longer need or use moral judgements. Guiding the reader through the diverse approaches to the complex subject, she points out the strong confident beliefs about such things as the value of freedom that underlie our supposed scepticism about values. She shows how the question of whether or not we can make moral judgements must inevitably affect our attitudes not only to the law and its institutions, but also to events that occur in our daily lives, and suggests that mistrust of moral judgement may be making life even harder for us than it would be otherwise.

Pbk ISBN 1-85399-166-X £6.95

Mind Matters series

DOES GOD EXIST?

Mark Corner

Do people talking about God know what they are talking about? Are they all talking about the same thing? How do different religions approach the existence of God? Can God's existence be proved? And even if it can, is God necessarily good?

Does God Exist? sets out to provide a lively and readable introduction to the main issues of theism and atheism. The author takes a fresh look at the question that has always been at the very roots of philosophy. His arguments provoke further interest as a source of new ideas.

Pbk ISBN 1-85399-164-3 £6.95

DO WE HAVE FREE WILL?

Mark Thornton

'Do We Have Free Will?' is not just an abstract philosophical question. It shapes educational and legal theory, it underlies particular views in Sociology and Psychology, and it is apparent in decisions in social and political policy.

Mark Thornton has produced a comprehensive guide to the rich variety of philosophical opinion on the subject. His clear exposition, penetrating criticism and original suggestions make this a valuable book for anyone with a practical or theoretical concern with issues of human freedom and responsibility.

Hbk ISBN 1-85399-018-3 £17.95
Pbk ISBN 1-85399-019-1 £6.95

MAD OR BAD?

Michael Bavidge

Michael Bavidge's question, 'Mad or Bad?' arises out of a long-standing philosophical interest in the connections between ethics and the philosophy of mind and from surprise at the verdicts in some notorious murder trials. Dreadful crimes demand total, unreserved condemnation and heavy punishment, but their very dreadfulness also leads us to think, 'Anyone who does *that* must be mad!'

This book provides a thought-provoking and sensitive new attempt to show us how we can preserve our ordinary moral intuitions about dreadful crimes while facing up to the difficulties of holding psychopathic criminals fully responsible for their actions. It is an important topic with wide-ranging implications which affect us all.

Hbk ISBN 1-85399-016-7 £17.95
Pbk ISBN 1-85399-017-5 £6.95

MINDS, BRAINS AND MACHINES

Geoffrey Brown

Whether or not machines can think is the stuff of which dreams and nightmares are made. Geoffrey Brown lives in this world, however, and his expertise in computer engineering gives him a down-to-earth view of the present complexity and future possibilities of the machines which appear to control much of our lives.

But he is also a philosopher who sees that the question demands careful consideration of some fundamental issues. What do we mean by 'think'? Is machine 'thinking' like human 'thinking'? What does it mean to be conscious? And in answering these questions we reach the heart of what is called the philosophy of mind.

Hbk ISBN 1-85399-012-4 £17.95
Pbk ISBN 1-85399-013-2 £6.95

REASONABLE CARE

Grant Gillett

If any discipline poses questions of life and death it is medical ethics. Where the availability of treatment is limited, who should get priority? Should doctors force treatment on unwilling patients? How can we judge whether a life is worth living and who should decide? *Reasonable Care* helps us approach these questions by offering a survey and critique of contemporary medical ethics, but it does so by drawing heavily on actual practice and the problems faced by doctors, nurses and their patients.

Grant Gillett is ideally placed to straddle the philosophical and the practical worlds, because when he is not writing and teaching philosophy he spends his time doing neurosurgery.

Hbk ISBN 1-85399-072-8 £17.95
Pbk ISBN 1-85399-073-6 £6.95